Math Mammoth Grade 4 Skills Review Workbook

By Maria Miller

Contents

Chapter 4: Time and Measuring

Chapter 5: Division

Foreword

Math Mammoth Grade 4 Skills Review Workbook has been created to complement the lessons in *Math Mammoth Grade 4* complete curriculum. It gives the students practice in reviewing what they have already studied, so the concepts and skills will become more established in their memory.

These review worksheets are designed to provide a spiral review of the concepts in the curriculum. This means that after a concept or skill has been studied in the main curriculum, it is then reviewed repeatedly over time in several different worksheets of this book.

This book is divided into chapters, according to the corresponding chapters in the *Math Mammoth Grade 4* curriculum. You can choose exactly when to use the worksheets within the chapter, and how many of them to use. Not all students need all of these worksheets to help them keep their math skills fresh, so please vary the amount of worksheets you assign your student(s) according to their needs.

Each worksheet is designed to be one page, and includes a variety of exercises in a fun way without becoming too long and tedious. We have created a spreadsheet document that lists the lessons spiraled in each worksheet. This document is included with the digital (download) version. You can also download it at the following link:

https://www.mathmammoth.com/skills_review_workbooks/guides/Skills_Review_Grade4_Spiraling_Guide.xls

The printed answer key can be purchased separately or in the digital download version it is included in the zip file.

I wish you success in teaching math!

Maria Miller, the author

Skills Review 1

1. Add mentally.

| a. $75 + 85 =$ _____ | b. $43 + 38 =$ _____ | c. $94 + 29 =$ _____ |

2. Starting at the top, find your way through the maze by coloring the number that is **one half** of the previous number.

1,250	1,410	1,840	1,120	1,680
640	710	930	840	860
310	230	460	540	420
120	160	240	190	210
64	85	122	105	100

3. Subtract from whole hundreds.

a. $300 - 4 =$ _____

b. $700 - 6 =$ _____

c. $900 - 9 =$ _____

4. Damian and his two brothers shared unequally the cost of a used car. Damian paid $2,715, Lyle paid $1,847, and Steve paid $2,069. How much did the car cost?

5. a. Carol baked 72 cookies. She gave some of them to Mrs. Harrison and now she has 48 left. How many cookies did she give to Mrs. Harrison? Use mental math.

b. The next morning, Carol counted the cookies and there were only 29 left. How many cookies had been eaten? Use mental math.

Puzzle Corner

What numbers are missing?

```
  5 □ 3 □              2 6 □ □
+ □ 9 □ 6           + □ □ 7 8
---------           ---------
  8 6 0 2             4 0 5 4
```

Skills Review 2

1. Subtract in columns. Check by adding!

a.	b.	c.
$\begin{array}{r} 7\ 0\ 5 \\ -\ 4\ 6\ 9 \\ \hline \end{array}$ $\quad +\ 4\ 6\ 9$	$\begin{array}{r} 9\ 2\ 5\ 0 \\ -\ 5\ 8\ 3\ 5 \\ \hline \end{array}$ $\quad +$	$\begin{array}{r} 6\ 0\ 1\ 3 \\ -\ 7\ 4\ 1 \\ \hline \end{array}$ $\quad +$

2. Alicia has $240. Does she have enough money
   to buy a bike for $190 and a helmet for $45?

   If yes, how much money does she have left?

   If no, how much more money does she need?

3. Add in any order, and in parts. Use mental math.

a. $85 + 5 + 9 + 70 + 8 + 32$	b. $216 + 90 + 7 + 3 + 4 + 300$

4. Compare. Write $<$, $>$, or $=$ in the box.

   a. $750 + 400$ ☐ $1400 - 250$ b. $7100 - 300$ ☐ $5800 + 800$

   c. $2300 - 800$ ☐ $900 + 500$ d. $920 + 400$ ☐ $2020 - 600$

5. Multiply.

a. $8 \times 2 =$ ____	b. $6 \times 1 =$ ____	c. $7 \times 8 =$ ____	d. $2 \times 0 =$ ____
$9 \times 9 =$ ____	$11 \times 10 =$ ____	$3 \times 4 =$ ____	$10 \times 6 =$ ____
$12 \times 5 =$ ____	$5 \times 3 =$ ____	$4 \times 12 =$ ____	$5 \times 11 =$ ____

Skills Review 3

1. The chart on the right shows the distances between a few Asian cities in kilometers.

   Mr. Wang is a businessman living in Hong Kong. One week, his job required him to travel from Hong Kong to Beijing, from Beijing to Tokyo, and then from Tokyo to Bangkok. How many kilometers did he travel in total?

	Beijing	Bangkok	Tokyo
Hong Kong	1,958	1,727	2,878
Beijing		3,294	2,091
Bangkok			4,600

2. Find the missing numbers. The sum of any two adjacent (side-by-side) numbers is the number directly above them.

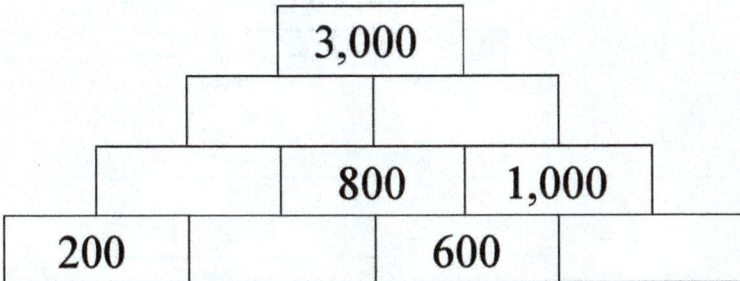

	3,000	
	800	1,000
200		600

3. Divide.

   a. $28 \div 7 =$ _____

   b. $45 \div 9 =$ _____

   c. $30 \div 5 =$ _____

   d. $72 \div 12 =$ _____

4. Figure out the pattern and continue it.

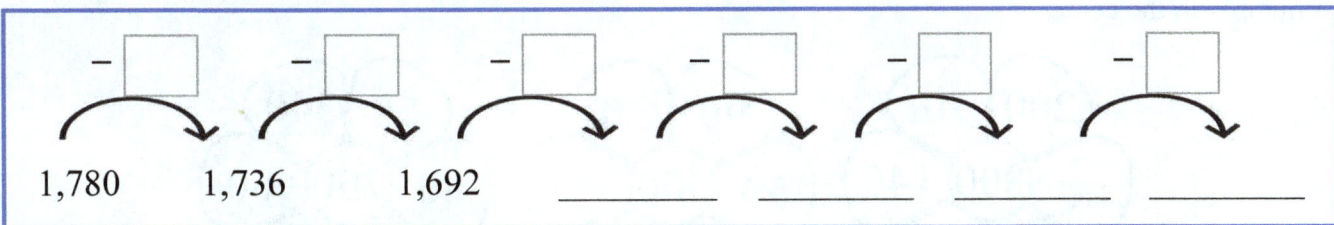

   1,780   1,736   1,692   _____   _____   _____

5. Farmer John has 1,348 sheep, separated into three flocks. One of the flocks has 430 sheep and another has 508 sheep. How many sheep does the third flock have?

Skills Review 4

1. First subtract, and then check by adding.

a. $7{,}100 - 4{,}820 - 695$	Add to check:
b. $3{,}092 - 463 - 78$	Add to check:

2. Write a missing addend problem that matches the bar model. Then solve it by subtracting.

a. 732, 244, x

_____ + _____ = _____

$x =$ _____ − _____ = _____

b. 945, x, 487

_____ + _____ = _____

$x =$ _____ − _____ = _____

3. Find the missing numbers. The sum of the petals on each flower should equal the number in the center.

Flower 1: 200, 70, 800 (center), 140, 210, 50

Flower 2: 90, 30, 50, 500 (center), 220, 80

Flower 3: 50, 340, 80, 700 (center), 60, 20

4. Write each number as a sum of its parts: thousands, hundreds, tens, and ones.

a. $7{,}948 =$	**b.** $3{,}092 =$

Skills Review 5

1. Calculate in the right order.

a. $6 \times 8 + 12 \div 3 =$ _____	**b.** $3 \times (11 - 3) \div 4 =$ _____	**c.** $90 - 7 \times 7 =$ _____

2. Add or subtract mentally in parts.

a.	**b.**	**c.**
$960 + 350 =$ _____	$2,000 - 406 =$ _____	$700 - 43 =$ _____

3. First, fill in the top row, continuing the pattern it has. Then add 69 to each number in the top row to get the number in the bottom row. *Hint: Instead of adding 69, add _____ , and then subtract __!*

n	730	690	650	610	570			
$n + 69$								

4. Solve. Write a number sentence for each problem. Do not just write the answer.

a. Clayton had 54 cherries. He and five of his friends shared them equally. How many did each child get?

b. Clayton and his friends EACH ate four cherries and saved the rest for later. In total, how many cherries did the children save for later?

5. Divide.

a. $100 \div 10 =$ _____	**b.** $36 \div 4 =$ _____	**c.** $84 \div 7 =$ _____

Puzzle Corner

Emma was organizing rows of chairs for a seminar. She arranged nine chairs in each of three rows and eight chairs in each of seven rows. Then, her boss asked her to remove one chair from each row. How many chairs were left? Write a number sentence.

Skills Review 6

1. Subtract in columns. Check by adding!

```
  7 0 3
− 4 9 5        +
_____     _____
```

Fill in the table.

×	4	7	9	6	8	12
6						
8						
4						
9						

3. Karen went on a nature hike. Each time she saw a bird, she wrote down what kind it was.
   Using her list (below), complete the frequency chart and make a bar graph.

dove dove dove dove dove dove sparrow sparrow sparrow sparrow sparrow sparrow
sparrow sparrow crow crow crow crow hawk hawk robin robin robin robin robin

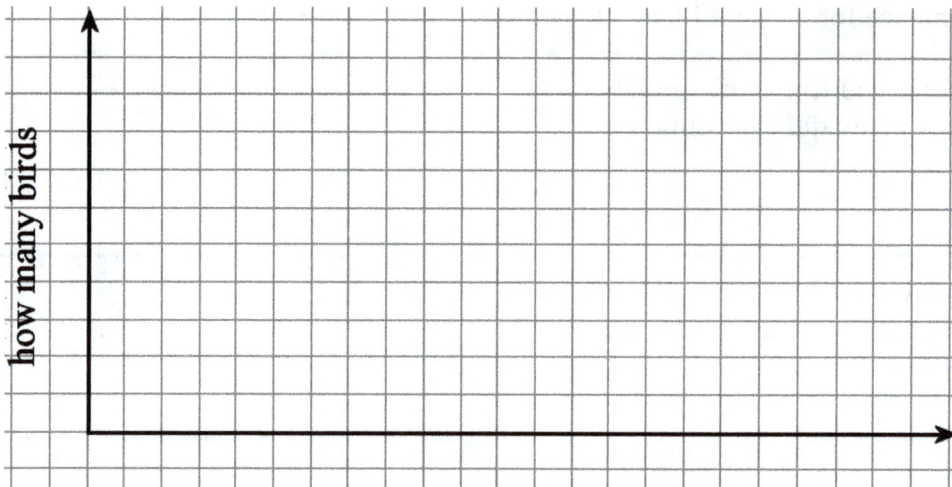

how many birds

Bird	Frequency
dove	
sparrow	
crow	
hawk	
robin	

Puzzle Corner

What numbers can go into the puzzle?

3,780	−		+		= 3,810
−		+		−	
	+		+		= 120
+		−		+	
	+		+		= 200
=		=		=	
3,750		10		150	

Skills Review 7

1. Circle the number sentence that fits the problem. Then solve for x.

a. Alex bought a bike for $95 and now he has $238 left. $x - 95 = 238$ OR $238 - 95 = x$ $x =$ _____	**b.** Farmer Benson had 128 cows. Then he bought 33 more. $x + 33 = 128$ OR $128 + 33 = x$ $x =$ _____

2. A cruise ship has 4,215 people on board. They are passengers, crew members, and two doctors. If there are 1,346 crew members, how many passengers are there?

3. The line graph shows how much money Becky earned each month walking her neighbors' dogs.

   a. During which month did Becky earn the most money?

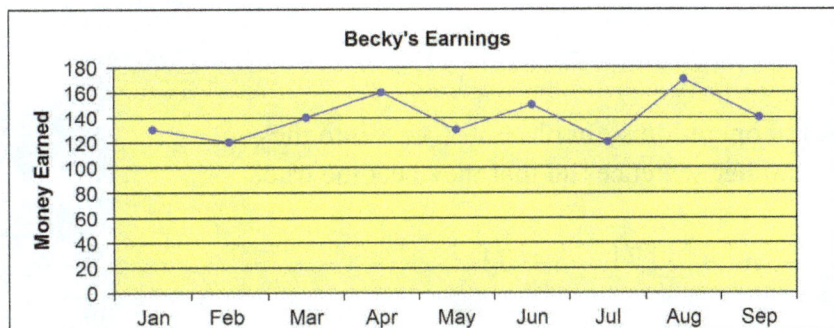

Becky's Earnings

   b. How much money in total did Becky earn during the months of February, March, and April?

4. Megan tried to complete this pattern, but she got confused! Find her mistakes and correct them.

n	330	370	410	450	490	530	570	610
$n + 79$	409	449	489	539	569	619	659	689

5. Add. Write the numbers under each other, carefully aligning the ones, tens, hundreds, and thousands. You may use a separate piece of paper if you prefer.

$6,830 + 1,597 + 305 + 28$

Skills Review 8

1. Round these numbers to the nearest ten, nearest hundred, and nearest thousand.

n	95	4,762	344
rounded to nearest 10			
rounded to nearest 100			
rounded to nearest 1000			

2. Subtract and compare the results. The problems are "related" — can you see how?

$14 - 6 = $ _____

$64 - 6 = $ _____

$142 - 60 = $ _____

$1,430 - 600 = $ _____

3. Write four different subtraction problems that are "related" to the problem $12 - 7 = 5$. See the examples above in exercise 2!

4. Put operation symbols $+$, $-$, or $\times$ into the number sentences so that they become true.

a. $9 \square 7 \square 8 = 71$

b. $8 \square (32 \square 29) \square 24 = 0$

c. $(9 \square 6) \square 4 = 60$

d. $6 \square 12 \square (41 \square 9) = 104$

5. Divide.

a. $42 \div 6 = $ _____

b. $100 \div 10 = $ _____

c. $36 \div 4 = $ _____

d. $56 \div 8 = $ _____

e. $15 \div 3 = $ _____

6. Solve $7,213 - 3,975 - 648$. Lastly, check by adding.

Skills Review 9

1. During a 45-minute period, Blake watched the cars going by his house and kept track of what color they were:

red blue red white green silver white red green blue white red silver white yellow red blue white blue red silver white red silver blue red green red yellow red white blue white green

Make a frequency table and finish the bar graph. (Note: Instead of writing the whole name of each color in the axis, you can just write the first letter to save space, if necessary.)

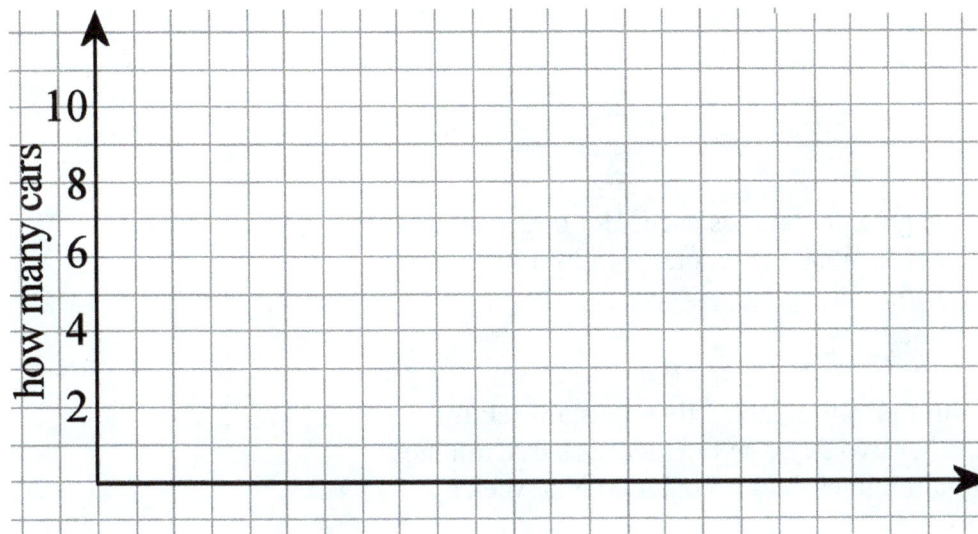

Color	Frequency

(bar graph with vertical axis labeled "how many cars" marked 2, 4, 6, 8, 10)

2. First estimate by rounding the numbers to the nearest hundred. Then find the exact answer.

Estimate:	Calculate exactly:
4,078 − 989 − 303 ↓ ↓ ↓ ≈ _____ − _____ − _____ = _____	

3. Stacy was traveling by car to visit her uncle. After driving 320 miles on the first day and 290 on the second day, she still had 865 miles left to drive.

   a. How many miles did Stacy have to drive to get to her uncle's house?

   b. How many miles did Stacy drive round-trip from her house to her uncle's?

Skills Review 10

1. How much is the discount, the new price, or the original price?

a.	**b.**	**c.**
Original $228.45 New price $219.95 Discount $_____	Original $24.96 New price _____ Discount $5.48	Original _____ New price $13.24 Discount $8.08

2. Solve the problems.

a. Anna has two cats that weigh eight pounds each. Her dog is six pounds heavier than either of the cats. What is the total weight of the animals?

b. Matthew practices his multiplication tables five days a week for ten minutes each day. The other two days, he practices for five minutes each day. How many total minutes does he practice in a week?

3. Make a word problem that matches the model. Then solve for x.

3,830	x

$\longleftarrow$ 6,750 $\longrightarrow$

$x =$ _____

4. First, fill in the top row, continuing the pattern it has. Then, find what number is being subtracted in the second row and finish subtracting.

n	360	440	520	600				
$n -$ ____		381		541				

Skills Review 11

1. Subtract in columns. Check by adding!

a.	b.	c.
$\begin{array}{r} 3\ 7\ 0\ 2 \\ -\ 3\ 5\ 0\ 4 \end{array}$ + _____	$\begin{array}{r} 2\ 0\ 6\ 1 \\ -\ 9\ 7\ 6 \end{array}$ + _____	$\begin{array}{r} 5\ 4\ 0 \\ -\ 1\ 8\ 3 \end{array}$ + _____

2. Solve.

a. _____ $- 70 = 365$	b. $90 +$ _____ $= 130$	c. $748 -$ _____ $= 688$

3. Find the double of the given numbers.

Number	70	230	350	800	2,700	5,500
Double the number						

4. Round these numbers to the nearest thousand.

a. $723 \approx$ _____   b. $49 \approx$ _____   c. $3,837 \approx$ _____

Puzzle Corner   What numbers can go into the puzzles?

	×		= 42
×		×	
	×		= 36
=		=	
63		24	

	×		= 21
×		×	
	×		= 40
=		=	
15		56	

Skills Review 12

1. Starting at the top and using a different color for each one, color three paths from top to bottom where the sum of the numbers is 2,100.

400	200	700	150	500
300	600	900	400	550
900	750	300	600	200
400	100	450	500	800
250	300	600	750	300

2. Subtract in your head.

a. $500 - 72 =$ _____

b. $483 - 99 =$ _____

c. $740 - 260 =$ _____

d. $900 - 45 =$ _____

3. Solve these problems with estimation. You don't need to find the exact answer!

a. Chad has $23.62 and he wants to buy a pair of shoes that cost $48.36. About how much more money does he need?

b. A pair of sunglasses costs $15.95 and a hat costs $9.26. About how much do they cost together?

4. Divide.

a. $72 \div 6 =$ _____

b. $108 \div 12 =$ _____

c. $40 \div 5 =$ _____

5. Jim planted 6 rows of lettuce. Each row is 10 feet long. He also has tomato plants in a row 4 feet long. Jim wants to weed the entire garden every week.

a. Make a schedule for him where he weeds about the same amount each day.

b. Make another plan in which he does not weed on Sunday.

6. Calculate in the right order.

a. $4 \times (9 - 3) =$ _____

b. $7 \times 7 + 12 \div 4 =$ _____

c. $(8 + 4) \times 6 + 7 =$ _____

Skills Review 13

1. Write the value of the underlined digit.

   a. 3,<u>6</u>09

   b. <u>8</u>,739

   c. 5,24<u>3</u>

   d. 4,<u>9</u>25

2. Add in columns.

   a.
   ```
        3 1 8
        7 4 5
          6 4
      5 9 0 7
        6 3 6
      2 1 0 3
   +    4 2 9
   ```

   b.
   ```
      4 0 8 2
      3 7 5 3
      1 8 0 5
        2 6 3
      1 6 2 9
        8 0 5
   +      7 8
   ```

3. Shauna kept track of how many cucumbers she harvested over an eight-week period. The table shows the results of her harvest.

   a. Make a line graph. Two values are already done for you.

   b. How many cucumbers did Shauna harvest in total?

Week	Cucumbers
1	12
2	18
3	22
4	36
5	44
6	40
7	30
8	24

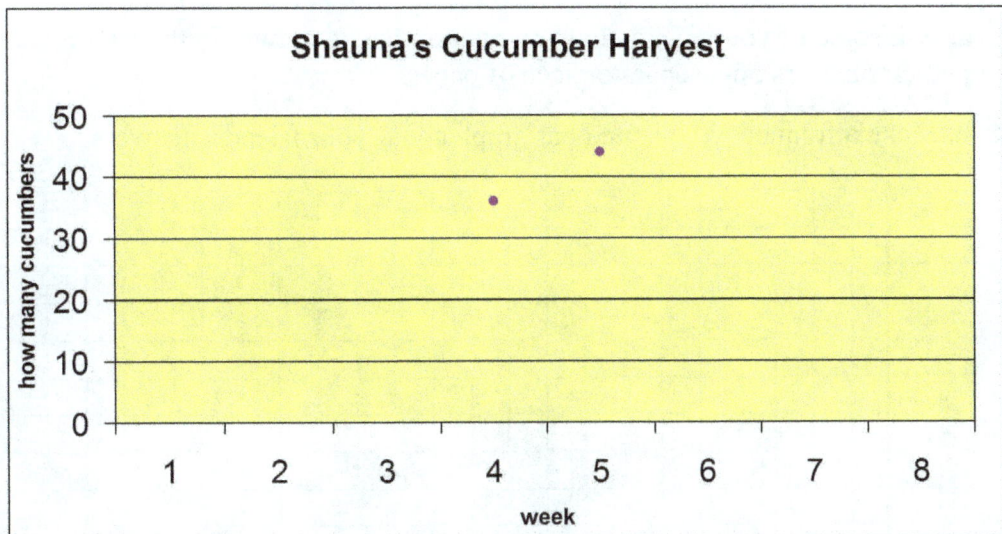

Shauna's Cucumber Harvest

4. Write a number sentence with x. Solve it.

Mr. Griffith invited 314 people to a seminar, but only 190 people attended. How many people didn't attend?

_____ − _____ = _____

$x =$ _____

Skills Review 14

1. Complete the *next* whole thousand.

a.	b.	c.
6,380 + _____ = _____	9,730 + _____ = _____	7,520 + _____ = _____

2. A package of six apples costs $9. How many apples can Alicia buy with $30?

3. Find the missing factors.

a. _____ × 4 = 48	b. _____ × 9 = 54	c. _____ × 7 = 42	d. _____ × 12 = 36

4. The table lists the number of miles a businessman traveled by airplane over a four-day period. Round the numbers to the nearest ten, and then estimate the total number of miles over these four days.

Day 1	Day 2	Day 3	Day 4
533	628	415	582

5. **a.** Ask some of your friends how many members there are in their family and write down their answers on a separate piece of paper.

b. Make a frequency table and bar graph about your friends' answers.

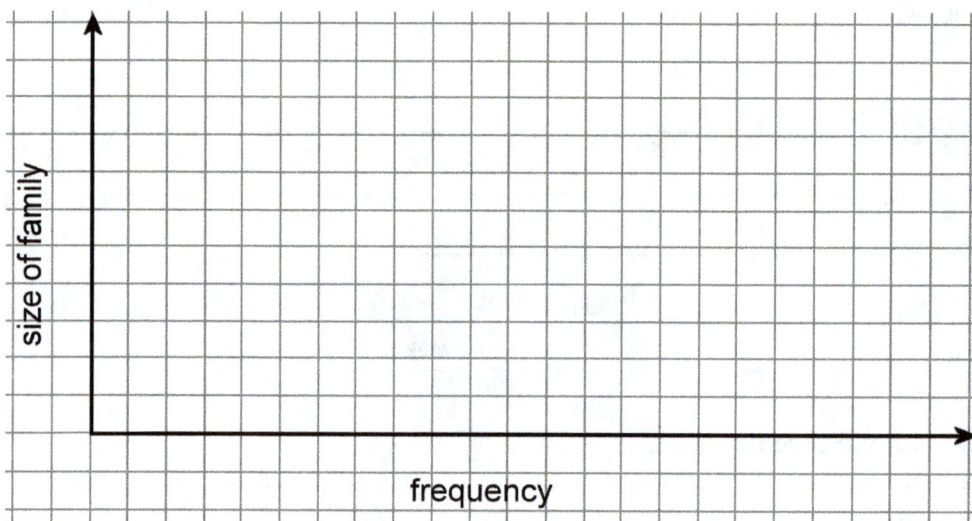

Size of family	Frequency

c. What was the most common size of family?   How many friends had that size of family?

Skills Review 15

1. Calculate in the right order.

a. $4 \times (7 - 5) + 9 =$ _____	**b.** $(8 + 4) \times 5 - 2 =$ _____	**c.** $9 \times 2 + 6 \div 3 =$ _____

2. Add and subtract, thinking in whole thousands.

a. $27{,}000 - 8{,}000 =$

b. $513{,}000 + 20{,}000 =$

c. $630{,}000 + 90{,}000 =$

d. $900{,}000 - 9{,}000 =$

3. Make change. Mark how many of each bill or coin you need.

Item cost	Money given	Change needed	$5 bill	$1 bill	25¢	10¢	5¢	1¢
a. $11.56	$20							
b. $8.33	$10							
c. $3.44	$5							
d. $13.97	$20							

4. Divide.

a. $21 \div 7 =$ _____	**b.** $48 \div 4 =$ _____	**c.** $72 \div 9 =$ _____
d. $60 \div 10 =$ _____	**e.** $30 \div 6 =$ _____	**f.** $16 \div 2 =$ _____

5. Sharon has 243 stamps in her stamp collection, and Rodney has 58 more stamps than Sharon. How many stamps do the two have in total?

6. Find the missing numbers in the additions.

a. $63 + 9 + 4 + 50 + 2 +$ _____ $= 165$	**b.** $100 + 328 + 3 +$ _____ $+ 7 + 30 = 553$

Skills Review 16

1. Write the numbers in expanded form.

   a. 79,312

   b. 605,483

   c. 17,604

2. Subtract with money amounts. Check by adding!

a.	**b.**
$\begin{array}{r} \$94.00 \\ -\ 38.62 \end{array}$ + _____	$\begin{array}{r} \$800.00 \\ -\ 619.47 \end{array}$ + _____

3. A bakery did a survey to find out their customers' favorite flavors of pie. Make two questions about the bar graph, write them below, and solve them.

   a.

   b.

Favorite Flavors of Pie

Coconut
Chocolate
Custard
Pumpkin
Apple

0   50   100   150   200

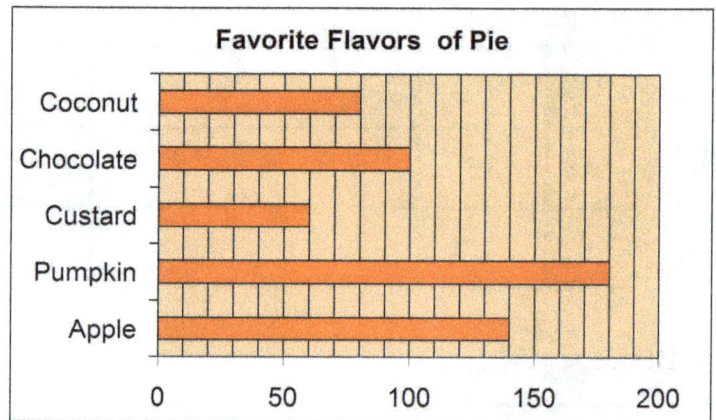

4. Round these numbers to the nearest ten, nearest hundred, and nearest thousand.

n	72	7,389	12,706
rounded to nearest 10			
rounded to nearest 100			
rounded to nearest 1000			

5. Add mentally.

   a. $63 + 82 =$ _____

   b. $46 + 98 =$ _____

   c. $34 + 27 =$ _____

Skills Review 17

1. Add the given numbers *and* the unknown *x* to the bar model. Then write an addition (a missing addend problem) and solve it.

Robert has 183 marbles. Some of them are white, 50 are blue, and 70 are green. How many marbles are white?

Addition:

Solution: $x =$ _____

2. Add or subtract the same number repeatedly.

a. Start at 5,700. Add 400 each time:	b. Start at 1,260. Subtract 80 each time:
_____	_____
_____	_____
_____	_____
_____	_____
_____	_____
_____	_____

3. Write < or > between the numbers.

a. 98,600 86,900	b. 13,040 13,400	c. 224,920 242,290
d. 59,060 59,006	e. 604,312 640,123	f. 450,083 45,830

4. Clarissa spends $157 a week on groceries, and Ashley spends $174.

   a. Round the numbers to the nearest ten, and estimate how much money each woman spends in four weeks.

   Clarissa: about _____

   Ashley: about _____

   b. What is the difference in the estimated amounts that the two women spend over the four-week period?

5. Subtract from whole thousands.

a. $8,000 - 25 =$ _____	b. $11,000 - 6 =$ _____	c. $2,000 - 30 =$ _____

Skills Review 18

1. Add or subtract.

a.	b.	c.
$\begin{array}{r} 702,458 \\ +79,650 \\ \hline \end{array}$	$\begin{array}{r} 800,000 \\ -648,091 \\ \hline \end{array}$	$\begin{array}{r} 234,370 \\ 110,000 \\ +376,859 \\ \hline \end{array}$

2. Make a number line from 450,000 to 470,000 with tick-marks at every whole thousand.
   Then mark the following numbers on the number line:
   456,000   463,000   452,000   467,000   460,000

3. Put operation symbols +, − , or × into the number sentences so that they become true.

a.	b.	c.
18 ☐ 9 ☐ 2 = 36	74 ☐ (12 ☐ 6) ☐ 2 = 0	8 ☐ 7 ☐ 9 = 47

4. The line graph shows how many people visited a small souvenir shop during a certain year.

   a. Approximately how many people visited the shop in April?

   b. In which months did the shop have between 100 and 150 visitors?

   c. In which months did the shop have at least 200 visitors?

   d. Approximately how many people visited the shop during the whole year?

Visitors to a Souvenir Shop

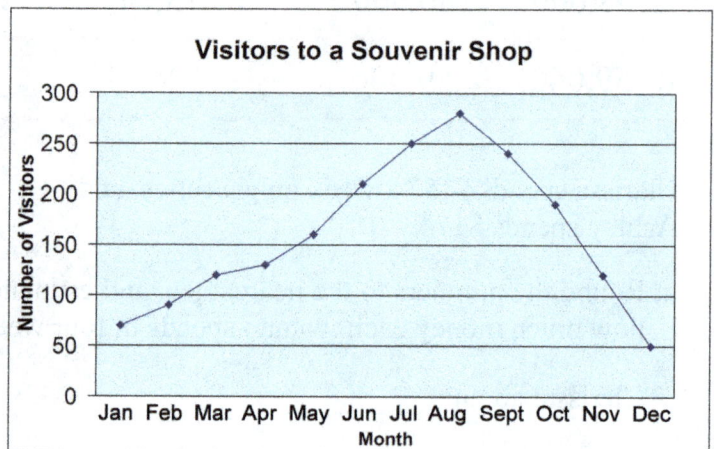

Skills Review 19

1. Round these numbers to the nearest thousand, nearest ten thousand, and nearest hundred thousand.

number	195,212	422,783
to the nearest 1,000		
to the nearest 10,000		
to the nearest 100,000		

2. Solve.

a. _____ $- 90 = 780$

b. $460 + 370 =$ _____

c. $920 -$ _____ $= 880$

d. $549 +$ _____ $= 619$

3. Multiply.

a. $6 \times 6 =$ _____	b. $4 \times 11 =$ _____	c. $9 \times 8 =$ _____
d. $12 \times 12 =$ _____	e. $3 \times 5 =$ _____	f. $0 \times 10 =$ _____

4. Solve. Write a number sentence with an unknown (x or ? or another symbol) for each problem. Then solve it.

a. Marsha bought a blouse for \$8.65, and she also bought a skirt. The total cost was \$18.50. How much did the skirt cost?	
b. Glenn bought a dictionary for \$7.35 and three notebooks for \$2.68 each. He had \$4.73 left. How much money did he have originally?	

5. First estimate by rounding the numbers to the nearest hundred. Then find the exact answer.

Estimate:	Calculate exactly:
$4,182 + 539 - 706$ ↓ ↓ ↓ $\approx$ ____ $+$ ____ $-$ ____ $=$ _____	

Skills Review 20

1. Figure out the pattern and continue it.

| + □ | + □ | + □ | + □ | + □ | + □ |

2360   2590   2820   _____   _____   _____   _____

2. Monica is reading 40 pages per day of a 320-page book. She has read for four days. How many pages does she have left to read?

3. Brian had $150. Then, he bought four shirts that cost $8 each, and three baseball caps that cost $7 each. How much money does he have left?

4. Subtract in columns. Check by adding!

a.	b.	c.
$\begin{array}{r} 8\ 6\ 2 \\ -\ 5\ 9\ 7 \end{array}$ +	$\begin{array}{r} 7\ 0\ 3\ 4 \\ -\ 4\ 2\ 4\ 6 \end{array}$ +	$\begin{array}{r} 1\ 2\ 2\ 0 \\ -\ \ 8\ 0\ 5 \end{array}$ +

5. **a.** What is the largest possible number you can build with the digits 3, 7, 5, and 1?

   b. What is the smallest possible number you can build with them?

   c. Subtract to find the difference between the largest possible number and the smallest possible number.

6. Divide.

a. $88 \div 8 =$ _____	b. $12 \div 3 =$ _____	c. $36 \div 9 =$ _____
d. $45 \div 5 =$ _____	e. $28 \div 7 =$ _____	f. $60 \div 12 =$ _____

Skills Review 21

1. Find the missing numbers. The sum of any two adjacent (side-by-side) numbers is the number directly above them.

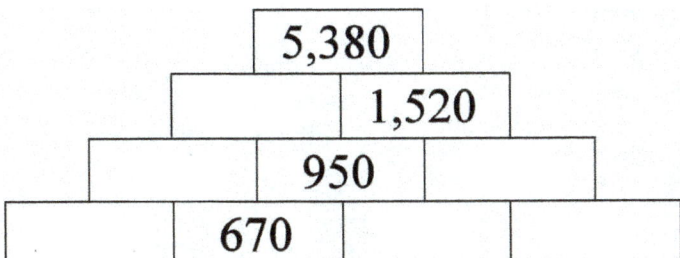

```
        5,380
           1,520
        950
     670
```

2. Subtract from whole thousands.

   a. $3{,}000 - 40 = $ _____

   b. $7{,}000 - 8 = $ _____

   c. $2{,}000 - 70 = $ _____

   d. $9{,}000 - 25 = $ _____

3. Jill asked some people how many servings of vegetables they eat per day. Below are the answers:
   9 7 4 2 9 5 8 4 7 8 5 4 9 1 7 3 6 8 2 7 9 5 8 4 7 6 7 1 4 7 2 9

   a. Make a frequency table and a bar graph.

   b. How many people ate 5 or less servings per day?

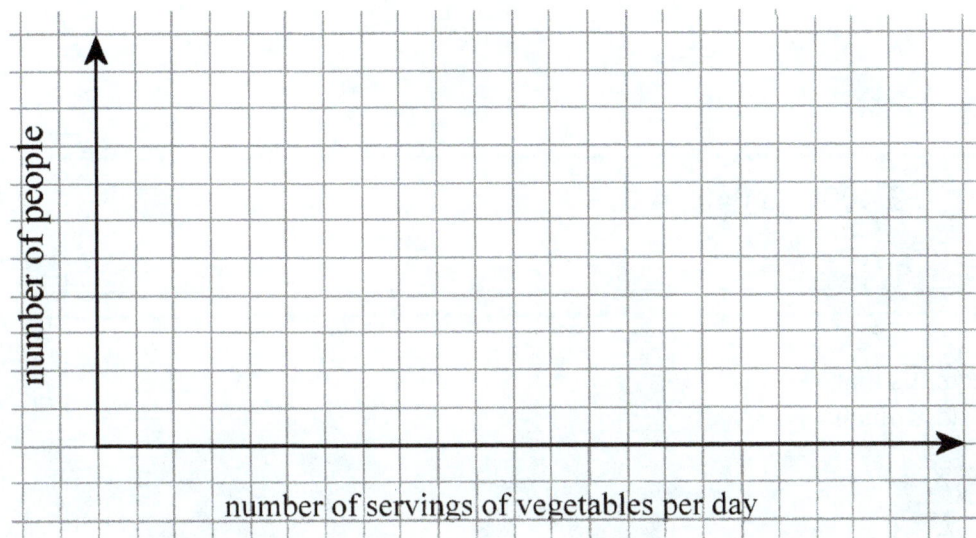

Servings per day	Number of people

number of people

number of servings of vegetables per day

4. There are 5,321 people in Cinnamon City. One August, the number of people increased to about 14,000 because of an art festival in the city.

   a. Did the number of people more than double?

   b. If so, how many more than double?
   If not, how many more visitors would be needed to double the original number of people in Cinnamon City?

Skills Review 22

1. Add or subtract.

a.	b.	c.
$\begin{array}{r} 902,045 \\ -\ \ 36,179 \\ \hline \end{array}$	$\begin{array}{r} 700,000 \\ -\ 485,627 \\ \hline \end{array}$	$\begin{array}{r} 528,700 \\ 194,080 \\ +\ 201,975 \\ \hline \end{array}$

2. Multiply.

a. $714 \times 100 =$ _____	b. $10 \times 2,730 =$ _____	c. $899 \times 1,000 =$ _____

3. Solve: $x + 943 = 1,590$.

4. Write the numbers in order.

430   4,300   4,030   3,400   3,040   4,003
_____ < _____ < _____ < _____ < _____ < _____

5. Caleb bought three volleyballs that cost $43.99 each. *About* how much was the total cost?

6. Divide.

a. $5,300 \div 100 =$ _____	b. $12,000 \div 10 =$ _____	c. $674,000 \div 1,000 =$ _____

7. Fill in the table, adding 1,999 each time. *Hint: First add _____, instead of _____. Then correct your answer.*

n	238	572	891	1,260	1,647	2,033
$n + 1,999$						

Skills Review 23

1. Write a number sentence for each of these problems. Use several operations in it. Then solve it.

Problem:	Number sentence:
a. Kim and Heather each picked seven flowers, and Nicky and Megan each picked eight flowers. How many flowers did they pick in total?	
b. Clara bought nine packages of cookies. Each package had six cookies. She gave four cookies to each of her three friends, and she ate three. How many cookies did she have left?	

2. Round these numbers to the nearest dollar.

    a. $3,624.50 ≈ _____   **b.** $92.49 ≈ _____   **c.** $706.28 ≈ _____

3. Write a subtraction problem that matches the bar model. Then solve it.

a.
$$\vdash\!\!\!-\!\!\!- X -\!\!\!-\!\!\!\dashv$$

280	824

_____ − _____ = _____

$x =$

b.
$$\vdash\!\!\!-\!\!\!- 1{,}048 -\!\!\!-\!\!\!\dashv$$

X	490

_____ − _____ = _____

$x =$

4. The bar graph shows how many points some children got while playing a computer game.

    a. What was the difference in the number of points that the player in second place got and the number of points that the player in third place got?

    b. How many more points would Marcella have needed in order to exceed Trent's points and win the game?

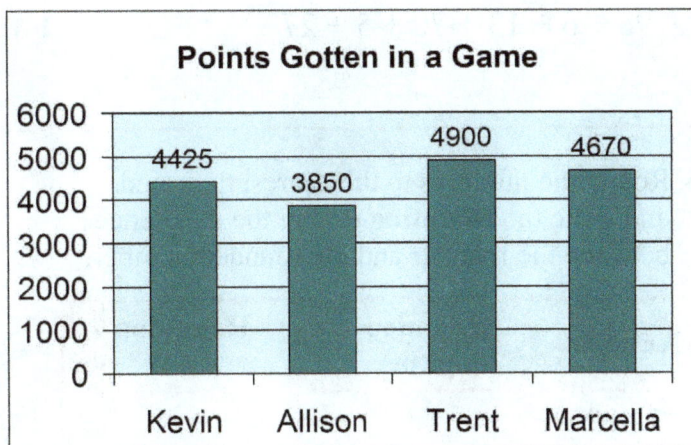

Points Gotten in a Game

	Points
Kevin	4425
Allison	3850
Trent	4900
Marcella	4670

Skills Review 24

1. Find the missing factors.

a. _____ × 6 = 36 _____ × 9 = 54	**b.** _____ × 3 = 27 _____ × 12 = 60	**c.** _____ × 8 = 96 _____ × 11 = 44	**d.** _____ × 4 = 28 _____ × 10 = 80

2. Robert, Eric, and Alex shared unequally the cost of a golf cart that cost $4,924. Alex paid $1,195 and Eric paid $1,428. How much did Robert pay?

3. Complete to the *next* whole thousand.

a.	b.	c.
4,670 + _____ = _____	2,420 + _____ = _____	9,780 + _____ = _____

4. Madeline picked 62 tomatoes and put them in bags, with 9 in each bag. She had some left over, so she used them to make salsa.

   a. How many bags of tomatoes did Madeline get?

   b. How many tomatoes did she use to make salsa?

5. Add in any order, and in parts. Use mental math.

a. 98 + 6 + 13 + 70 + 5 + 27	**b.** 142 + 49 + 7 + 4 + 11 + 800

6. Round the numbers to the nearest thousand, and write the **rounding error:** the difference between the number and the rounded number.

Number	Rounded number	Rounding error
7,489		
2,940		
5,720		

7. Subtract mentally.

a. 600 − 8 = _____
b. 340 − 70 = _____
c. 53 − 17 = _____
d. 2,000 − 19 = _____

Skills Review 25

1. Write an addition or a subtraction with an unknown (x or ?). Solve it.

A motorcycle used to cost $7,460, but then the store discounted it and the new price was $6,945. How much was the discount?

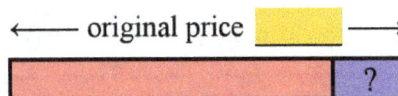

←—— original price ——→

?

_____

_____

2. What are these numbers?

   a. It has six digits. The ones digit is half of the hundred thousands digit, the thousands digit is double the tens digit, the tens digit is four and six is in the hundred thousands place. The digit sum is 21.

   b. All five digits of this number are odd numbers, and each is different. The ten thousands digit is three times the thousands digit, and the thousands digit is two less than the tens digit. The ones digit is seven.

3. Find the numbers that go on the empty lines.

   a. $70 - \underline{\qquad} = 48$ | **b.** $23 + 59 = 2 \times \underline{\qquad} + 2$ | **c.** $7 \times 5 - 6 = 4 \times \underline{\qquad} + 5$

4. Add. Write the numbers under each other, carefully aligning the ones, tens, hundreds, and thousands. You may use a separate piece of paper if you prefer.

   $347,846 + 62,503 + 92 + 208 + 4,725$

5. Seventy-six teachers attended a workshop. They were seated at tables for eight.

   a. How many tables were full?

   b. How many people were seated at the table that wasn't full?

6. People went shopping with different amounts of money. Complete the table.

Person	Jessica	Dawson	Vanessa
Amount of money to spend	$326	$413	
How much money was spent	$78		$46
How much money is left		$318	$158

Skills Review 26

1. Find the missing factor.

a. _____ × 8 = 480	**b.** 60 × _____ = 660	**c.** _____ × 30 = 2,700

2. Jeanine kept track of how many eggs she gathered during one week.

   a. Make a line graph. Remember, the one axis is "days" and the other is "how many eggs."
   Also, you need to decide the scaling for the vertical axis.

   b. How many eggs did Jeanine gather in total that week?

Day	Eggs
Mon	12
Tue	8
Wed	10
Thu	14
Fri	12
Sat	16
Sun	10

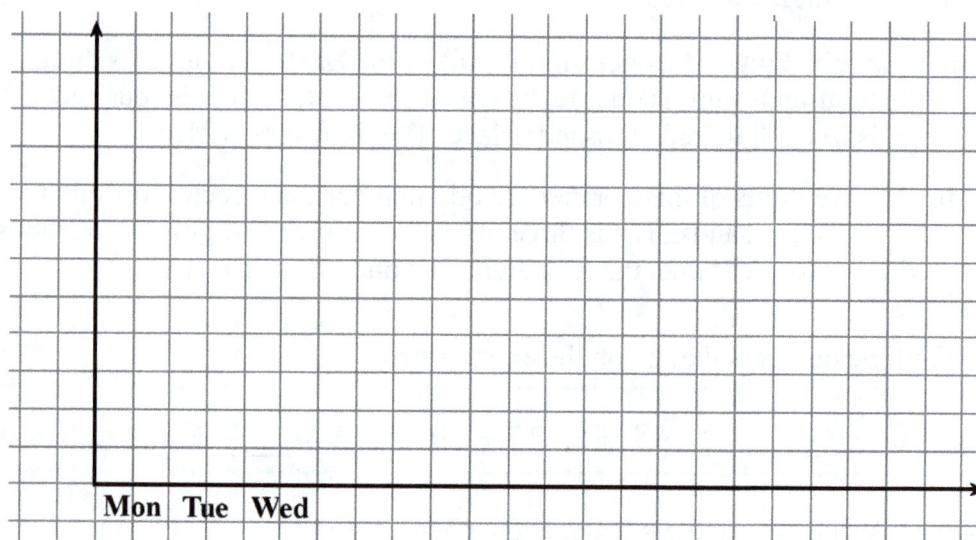

3. Use rounded numbers to solve these problems.

a. *Round the numbers to the nearest thousand.*	**b.** *Round the numbers to the nearest hundred.*
Alex pays $35,744 a year in rent. That means about $_____. So, he pays about _____ *monthly.*	A certain book had 429 pages with an average of 278 words on each page. That is about _____ words in the whole book.

4. Describe a shopping situation where you need to do these calculations:

   a. $28 + 3 × $4.35

   b. $36 − 3 × $2.68

Skills Review 27

1. Fill in the missing numbers. Write the area of the *whole* rectangle as a SUM of the areas of the *smaller* rectangles. Also find the total area.

a. $\underline{8}$ × $\underline{29}$ = ___ × _____ + ___ × ___

20, 9
8

= _____

b. ___ × _____ = ___ × _____ + ___ × ___

30, 7
9

= _____

2. Robert flew round-trip from Madrid to Jakarta. Candace flew from Sydney to Jakarta, and then from Jakarta to Quito.

   a. Who traveled more miles?

   b. How many more miles?

	Sydney, Australia	Quito, Ecuador	Jakarta, Indonesia
Madrid, Spain	17,674	8,732	12,181
Sydney, Australia		13,599	5,494
Quito, Ecuador			19,081

3. Write < or > between the numbers.

a. 61,305 63,501	**b.** 59,264 56,294	**c.** 75,885 758,850
d. 192,468 192,846	**e.** 263,309 623,390	**f.** 82,405 84,502

4. Figure out the pattern and continue it.

+ [ ]   + [ ]   + [ ]   + [ ]   + [ ]   + [ ]

5,600   6,200   6,800   _____   _____   _____   _____

Skills Review 28

1. Estimate the cost. Round one or both numbers so that you can easily multiply in your head.

a. 18 calendars at $7.35 per calendar	**b.** 12 cabinets at $338 each
≈	≈

2. Add or subtract, thinking in whole thousands.

   a. $78,000 - 9,000 =$

   b. $368,000 + 40,000 =$

   c. $670,000 + 30,000 =$

   d. $520,000 - 50,000 =$

3. Write using numbers.

a. 83 thousands _____	**b.** 66 tens _____	**c.** 45 hundreds _____

4. Divide.

a. $84 \div 12 =$ _____	**b.** $50 \div 5 =$ _____	**c.** $63 \div 7 =$ _____	**d.** $36 \div 3 =$ _____

5. Subtract.

a.	**b.**	**c.**
$\begin{array}{r} 370,056 \\ -\ 92,788 \\ \hline \end{array}$	$\begin{array}{r} 897,056 \\ -\ 43,298 \\ \hline \end{array}$	$\begin{array}{r} 824,000 \\ -\ 6,437 \\ \hline \end{array}$

Puzzle Corner

Mrs. Erickson asked her class of 26 students to each bring nine popsicle sticks to class the next day for a craft project. Some of her students were only able to bring seven popsicle sticks, so in total, there were 226. How many students only brought seven popsicle sticks?

Skills Review 29

1. Find the missing factors.

a. ____ × 2 = 18	b. ____ × 6 = 42	c. ____ × 4 = 20	d. ____ × 3 = 24
____ × 11 = 33	____ × 8 = 16	____ × 9 = 108	____ × 7 = 63

2. Circle the number sentence that fits the problem. Then solve for x.

a. Conrad had some marbles in a bag. Alex gave him 25 more, and now he has 68.	b. Lauren had $92. She bought a purse and now she has $74 left.
$x + 25 = 68$ OR $25 + 68 = x$	$92 - x = 74$ OR $x - 92 = 74$
$x =$ _____	$x =$ _____

3. Grace and five of her friends shared 57 peanuts as equally as possible, and they gave the rest to a pigeon. How many peanuts did the pigeon get?

4. Subtract from whole thousands.

a. $7,000 - 80 =$ _____	b. $4,000 - 400 =$ _____	c. $2,000 - 9 =$ _____

5. Solve these problems with estimation.

a. A refrigerator is $749 and a stove is $1,615. What is the total approximately?	b. A coffee cup costs $8.95. About how many cups can you buy with $40?

6. Subtract. Lastly, check by adding.

$8,030 - 3,960 - 823$	Add to check:

Skills Review 30

1. Leah, Bill, and Joy sold boxes of granola bars at a fund-raiser for their school. Each box contained 10 granola bars. Leah sold 32 boxes, Bill sold 45 boxes, and Joy sold 18 boxes. What was the total number of granola bars they sold?

2. Break the second factor into thousands, hundreds, tens, and ones. Multiply separately, and add.

a. $4 \times 2{,}953$	b. $9 \times 1{,}687$	c. $8 \times 3{,}024$
+ _____	+ _____	+ _____

3. Marsha wrote down the hair color of people who walked by her house during a two-hour period:

blond black brown gray blond red black black blond blond blond brown gray red gray blond black brown brown gray gray blond brown blond black red brown gray

a. Make a frequency table and a bar graph.

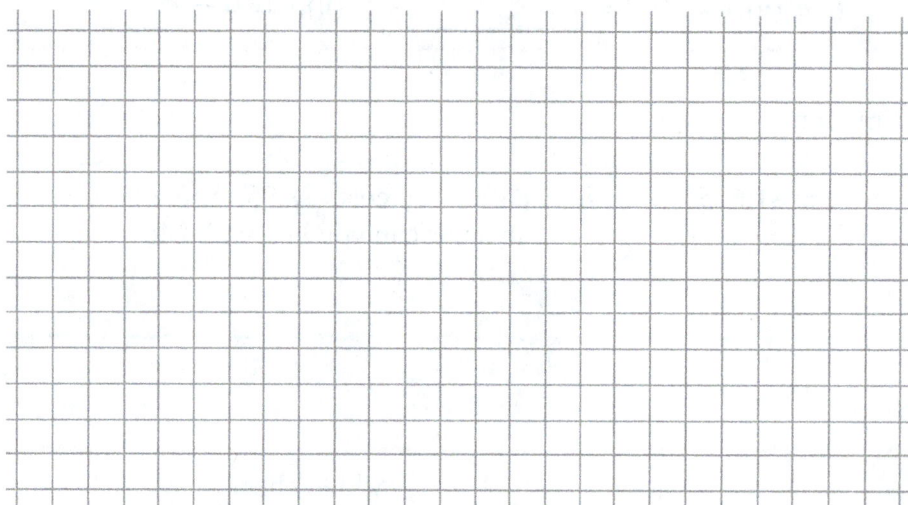

Hair Color	Frequency

b. What was the least common hair color?

c. How many people walked by during the two-hour period?

4. Fill in the empty lines with the correct numbers.

a. $138 + $ _____ $= 218$	b. $9 + 8 + 5 = $ _____ $- 30$	c. $7 \times 60 = 90 + $ _____

Skills Review 31

1. Multiply.

a. 50 × 5000 = _____	**b.** 20 × 80 = _____	**c.** 900 × 900 = _____
300 × 700 = _____	6000 × 400 = _____	60 × 2200 = _____

2. How much is the discount, the new price, or the original price?

a. Old price $175.85 New price $156.30 Discount $_____	**b.** Before $_____ Now $63.98 Discount $8.35

3. Write number sentences (additions, subtractions, multiplications) on the lines, and solve.

a. Sally, Brett, Aaron, and Megan each collected 37 seashells and James collected 20. How many seashells did they collect in total?

_____

b. Joshua earns $12 an hour. How many hours would he have to work in order to earn $192? Guess and check.

_____

_____

4. Round these numbers to the nearest thousand, nearest ten thousand, and nearest hundred thousand.

number	468,237	216,951	847,623	566,315
to the nearest 1,000				
to the nearest 10,000				
to the nearest 100,000				

Skills Review 32

1. Calculate in the right order.

$8236 + 7 \times 69 - 58$

2. Multiply. Be careful with the regrouping.

a.
$$\begin{array}{r} 9\ 7 \\ \times\quad 9 \\ \hline \end{array}$$

b.
$$\begin{array}{r} 4\ 3 \\ \times\quad 6 \\ \hline \end{array}$$

c.
$$\begin{array}{r} 6\ 7 \\ \times\quad 3 \\ \hline \end{array}$$

d.
$$\begin{array}{r} 5\ 8 \\ \times\quad 4 \\ \hline \end{array}$$

3. Solve for x.

a. $84 - x = 36$

$x =$ _____

b. $x - 28 = 69$

$x =$ _____

c. $926 - x = 472$

$x =$ _____

4. Write a number sentence and solve.

a. Karen picked some flowers to sell. She made seven bouquets of eight flowers each, and four bouquets of nine flowers each. Then, she gave the six remaining flowers to her mother. How many flowers did Karen pick?

b. Patty has 60 tomato plants. If she plants them in rows of seven, how many full rows will she have?

How many tomato plants will she have in the partial row?

Skills Review 33

1. Draw a bar model. Then fill in the missing parts in the number sentence and solve for x.

Seth bought six t-shirts for $14.45 each.
Now he has $79.68 left. How much did
he have originally (x)?

$$x = 6 \times \$\rule{1.5cm}{0.4pt} + \$\rule{1.5cm}{0.4pt}$$

2. Calculate. Line up all the place value units carefully.

    a. $23{,}040 - 7{,}562$ **b.** $61 + 542{,}039 + 748 + 3{,}575$

3. Estimate the results by rounding one or both factors. Don't round both numbers if you can multiply in your head just by rounding one factor.

a. 7×88	**b.** 35×243	**c.** 692×9
$\approx \rule{1cm}{0.4pt} \times \rule{1.5cm}{0.4pt} = \rule{1.5cm}{0.4pt}$	$\approx \rule{1cm}{0.4pt} \times \rule{1.5cm}{0.4pt} = \rule{1.5cm}{0.4pt}$	$\approx \rule{1cm}{0.4pt} \times \rule{1cm}{0.4pt} = \rule{1.5cm}{0.4pt}$

4. **a.** During which months does Paris get an average of at least 25 mm of rain?

    b. During which months does Paris get an average of less than 20 mm of rain?

    c. Does Paris have a clear and definite dry season sometime during the year?

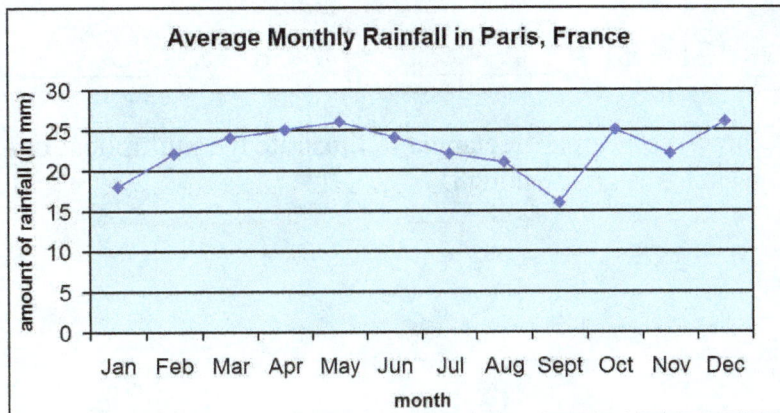

Average Monthly Rainfall in Paris, France

amount of rainfall (in mm)

30 · 25 · 20 · 15 · 10 · 5 · 0

Jan Feb Mar Apr May Jun Jul Aug Sept Oct Nov Dec
month

Skills Review 34

1. Do the numbers on the petals add up to the number in the center of the flower? If yes, color the flower. If no, cross it out, and write below it the difference between the sum of the numbers on the petals and the number in the center.

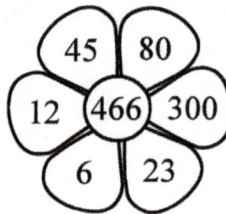

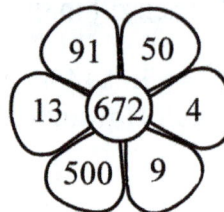

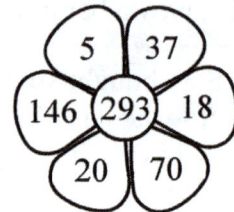

Flower 1: center 466, petals 45, 80, 300, 23, 6, 12

Flower 2: center 672, petals 91, 50, 4, 9, 500, 13

Flower 3: center 293, petals 5, 37, 18, 70, 20, 146

2. Write < or > between the numbers.

| **a.** 369,764   396,467 | **b.** 27,118   217,180 | **c.** 75,324   75,243 |

3. Solve.

| **a.** $540 \times \underline{\hspace{1cm}} = 54,000$ | **b.** $326 \times \underline{\hspace{1cm}} = 3,260$ | **c.** $\underline{\hspace{1cm}} \times 79 = 79,000$ |

4. Fill in the table, thinking logically! Mike can type 180 words in four minutes. Write in the table how many words he types in the given amount of minutes.

Words				180						
Minutes	1	2	3	4	5	6	7	8	9	10

5. Matthew bought a basketball for $12.65, a baseball cap for $6.79, and a package of socks for $7.48. He paid with $50. How much change did he receive?

6. Divide.

| **a.** $121 \div 11 = \underline{\hspace{1.5cm}}$ | **b.** $56 \div 8 = \underline{\hspace{1.5cm}}$ | **c.** $63 \div 7 = \underline{\hspace{1.5cm}}$ |

7. Draw a *three*-part rectangle to illustrate the multiplications. You don't have to draw accurately; a sketch is good enough.

9×284

$= \underline{\hspace{0.5cm}} \times \underline{\hspace{1cm}} + \underline{\hspace{0.5cm}} \times \underline{\hspace{0.5cm}} + \underline{\hspace{0.5cm}} \times \underline{\hspace{0.5cm}}$

$=$

Skills Review 35

1. Subtract with money amounts. Check by adding!

a.	b.
$\begin{array}{r} \$90.00 \\ -\ 42.86 \end{array}$ + _____	$\begin{array}{r} \$500.00 \\ -\ 391.75 \end{array}$ + _____

2. Mark has $595 and Todd has $738.
How much more money do they need
to buy a rowboat that costs $1,402?

3. Allison made some number patterns. Figure out what kind of patterns they are.

a.
560,000	280,000	140,000	70,000	35,000	17,500	8,750	4,375

b.
85	170	340	680	1,360	2,720	5,440	10,880

4. Write the numbers and x into the bar model.
Notice carefully which number is the *total*.
Then write a subtraction that helps you solve x.

$$147 - x = 68$$

$x =$ _____ $-$ _____ $=$ _____

5. Find the missing factors.

a. $6 \times$ _____ $= 72$

b. _____ $\times 9 = 81$

c. _____ $\times 6 = 48$

d. $11 \times$ _____ $= 0$

6. Fill in the missing numbers. Write the area of the *whole* rectangle as a SUM of the areas of the *smaller* rectangles. Also find the total area.

$37 \times 69 =$ _____ $\times$ _____ $+$ _____ $\times$ _____

$+$ _____ $\times$ _____ $+$ _____ $\times$ _____

$=$

Skills Review 36

1. Multiply. Place a zero in the ones place before multiplying.

a.	b.	c.	d.	e.
7 9	8 2	5 0 7	1 4 6	3 2 5
× 6 0	× 3 0	× 4 0	× 9 0	× 2 0
0				

2. Write an addition or a subtraction with an unknown (*x* or ?). Solve it.

A laptop used to cost $929, but now it is discounted and costs $885. How much is the discount?

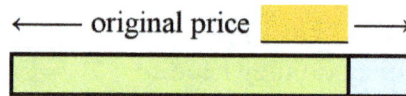

←—— original price ——→

_____

_____

3. Make a number line from 690,000 to 710,000 with tick-marks at every whole thousand.
   Then mark the following numbers on the number line:
   697,000   704,000   709,000   693,000   702,000

4. Fill in the table.

×	0	1	2	3	4	5	6	7	8	9	10	11	12
9													
4													
12													
7													
11													
8													
6													

Skills Review 37

1. Calculate.

a. $90 - (12 + 8) \times 3 =$ _____	**b.** $(3{,}000 - 900 - 400) \times 2 - 800 =$ _____

2. Starting at the top, find a way through the maze. The rule is: each quotient on your path must be one half of the previous quotient. Color the boxes in the path you follow.

$24 \div 3$	$21 \div 7$	$144 \div 12$	$33 \div 11$
$10 \div 2$	$12 \div 2$	$72 \div 3$	$4 \div 4$
$6 \div 2$	$66 \div 11$	$9 \div 3$	$36 \div 6$

3. List 3 numbers between 2 and 10 that can divide evenly into the given numbers.

a. 32
b. 28

4. Round the numbers to the nearest thousand, and write the **rounding error:** the difference between the number and the rounded number.

Number	Rounded number	Rounding error
7,634		
5,249		
3,768		

5. Multiply. Estimate the answer on the line.

a. 8×273	**b.** 6×594
$\approx$ _____	$\approx$ _____

6. Write a number sentence. Then solve the problem.

Mr. Erickson had $54,235 in savings. Then, he loaned $6,500 to his son Adam and twice that much to his son Brandon. How much money does he have left in his savings?

Skills Review 38

1. Multiply.

a.
```
      2 6
x     9 4
```

b.
```
      4 2
x     5 3
```

c.
```
      5 7
x     6 8
```

d.
```
      9 0
x     3 3
```

2. Solve the pan balance equations.

a. One triangle = _____

b. One square = _____

3. Greg and Taylor work for two different landscaping companies. Greg earns $523 a week, and Taylor earns $2,849 a month.

a. Estimate how much money each landscaper earns in two months.

Greg: about _____

Taylor: about _____

b. What is the difference in the two men's estimated earnings over a 2-month period?

4. Fill in.

a.

Days	Hours
5	
6	
7	

b.

Minutes	Seconds
6	
7	
8	

c.

Years	Months
9	
10	
11	

Skills Review 39

1. How much time passes? Figure it out in parts.

a. From 7:45 to 12:30	**b.** From 5:25 AM to 1:20 PM

2. Solve the problems using the tables. First find out the answer for ONE of the items.

a.

1 shirt	
7 shirts	
8 shirts	$48

b.

1 notebook	
5 notebooks	$2.50
12 notebooks	

3. Multiply.

a. $40 \times 70 =$ _____

b. $90 \times 600 =$ _____

c. $800 \times 300 =$ _____

4. Divide.

a. _____ $\div 7 = 12$	**d.** $63 \div$ _____ $= 7$
b. $108 \div$ _____ $= 9$	**e.** _____ $\div 7 = 20$
c. $56 \div 8 =$ _____	**f.** $48 \div 6 =$ _____

5. Solve.

a. It takes about 80 minutes to drive from Carl's home to the store. He is going to drive to the store and spend about 3 hours shopping, and then come back home. What is the total hours and minutes it will take for his shopping trip?

_____

b. Millie finished swimming laps in the pool in exactly two minutes, and Joan was 24 seconds faster. What was Joan's finishing time in seconds?

_____

Skills Review 40

1. Break the second factor into thousands, hundreds, tens, and ones. Multiply separately, and add.

a. $5 \times 8{,}172$	b. 3×946	c. $7 \times 5{,}283$
+ _____	+ _____	+ _____

2. Find each missing number so that the four numbers on the sides add up to the number in the middle.

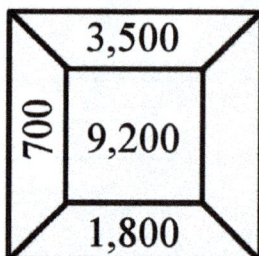

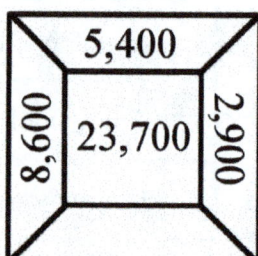

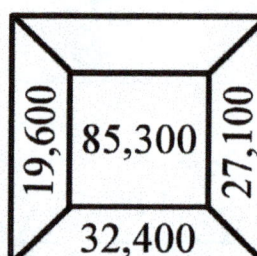

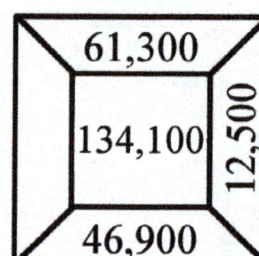

3,500	5,400		61,300
700 9,200 1,800	8,600 23,700 2,900	19,600 85,300 27,100 32,400	134,100 12,500 46,900

3. How much time passes? Use subtraction.

a.	b.	c.
From 3:33 p.m. till 7:19 p.m.	From 2:51 p.m. till 5:10 p.m.	From 7:48 a.m. till 11:24 a.m.
7 h 19 m − 3 h 33 m		

4. Clarissa has $72. How many packages of batteries can she buy that cost $6 each?

5. Estimate the cost. Round one or both numbers so that you can multiply in your head.

a. 250 toy cars at $3.78 each	b. 600 pounds of oranges at $1.22 per pound
≈	≈

Skills Review 41

1. The graph shows how much money the Cooper Family spent on groceries each month for six months.

   a. During which months did the Cooper Family spend less than $1,000 on groceries?

   b. *Estimate* the total amount they spent during the six months.

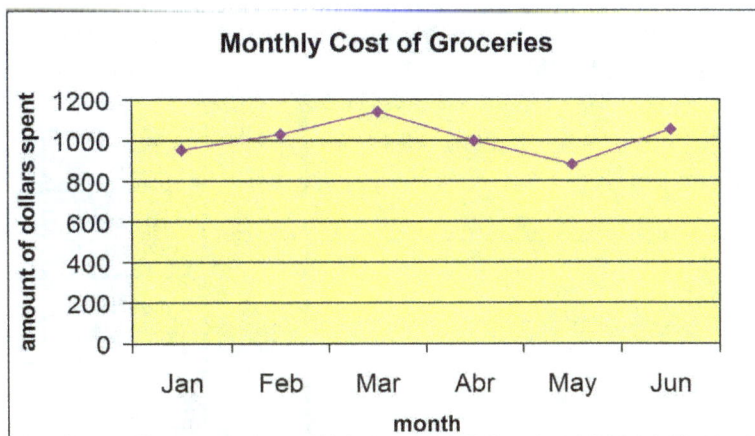

Monthly Cost of Groceries

amount of dollars spent

1200
1000
800
600
400
200
0

Jan Feb Mar Abr May Jun

month

2. Calculate.

   a. $748,602 + 36,975 + 843$

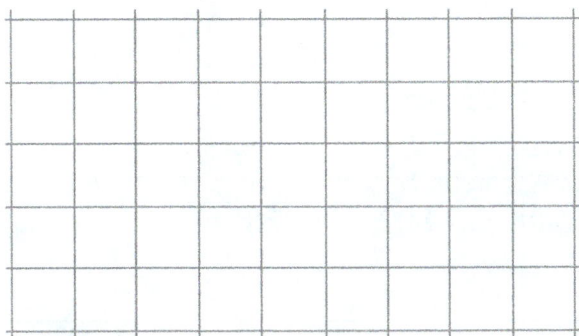

   b. $572,405 - 19,730$

3. Complete:

   a. The freezing point of water is _____°C.   **b.** The boiling point of water is _____°C.

   c. Normal body temperature is _____°C.   **d.** The temperature inside now is _____°C.

4. Calculate.

   a. $8 \times \$27.56$

   b. $4 \times \$83.14$

5. Drew practices playing his violin from 2:30 pm to 4:15 pm six days a week. How many hours does he practice in two weeks?

Skills Review 42

1. Skip-count backwards by 900s, starting at 8,600.

8,600

_____

_____

_____

_____

2. Round these numbers to the nearest ten, nearest hundred, and nearest thousand.

n	73	3,453	6,786
rounded to nearest 10			
rounded to nearest 100			
rounded to nearest 1000			

3. Last week, five of a factory's employees worked 53 hours, and seven employees worked 45 hours. How many hours did they work in total?

4. Write the numbers in order.

7,082   7,820   7,280   7,208   7,802   7,028

_____ < _____ < _____ < _____ < _____ < _____

5. Solve.

a. $9 \times$ ____ $= 54$

____ $\times 6 = 66$

b. ____ $\times 4 = 28$

$12 \times$ ____ $= 132$

c. $8 \times$ ____ $= 64$

____ $\times 4 = 48$

d. ____ $\times 11 = 121$

____ $\times 7 = 49$

6. Describe a situation to fit these temperatures.

   a. 95°F

   b. 26°F

7. Divide.

a. $70,000 \div 100 =$ _____

$585,000 \div 1,000 =$ _____

b. $4,530 \div 10 =$ _____

$8,600 \div 100 =$ _____

c. $90,000 \div 1,000 =$ _____

$62,000 \div 10 =$ _____

Skills Review 43

1. Change the 24-hour clock times to the a.m. / p.m. times.

a. 19:38	b. 6:50	c. 21:45	d. 13:29
_____ : _____ p.m.	_____ : _____	_____ : _____	_____ : _____

2. Divide.

a. $63 \div 9 =$ _____	b. $36 \div 6 =$ _____	c. $28 \div 4 =$ _____
d. $72 \div 12 =$ _____	e. $49 \div 7 =$ _____	f. $56 \div 8 =$ _____

3. Break the money amounts into dollars and cents. Multiply separately, and lastly add.

a. $7 \times \$9.40$ _____ + _____ = (7 × $9) (7 × $0.40)	b. $9 \times \$8.25$
c. $6 \times \$4.72$	d. $5 \times \$3.60$

4. Solve the problem.

> Karen spends 45 minutes each day walking her dog. How much time does she spend in two weeks? Give your answer in minutes, and also as hours and minutes.
>
> _____

5. Insert the given numbers *and* the unknown x into the bar model. Then write a missing addend problem and a subtraction to solve it.

> Kevin had $345. Then, after working one week for a neighbor, he had $520. How much did he earn that week?
>
> ← ——— total ———— →
>
> _____ + _____ = _____
>
> $x =$ _____ − _____ = _____

Skills Review 44

1. Multiply.

a.
$$\begin{array}{r} 3\ 9 \\ \times\ 5\ 0 \\ \hline \end{array}$$

b.
$$\begin{array}{r} 5\ 6 \\ \times\ 1\ 0 \\ \hline \end{array}$$

c.
$$\begin{array}{r} 7\ 3\ 4 \\ \times\ 9\ 0 \\ \hline \end{array}$$

d.
$$\begin{array}{r} 8\ 0\ 2 \\ \times\ 7\ 0 \\ \hline \end{array}$$

e.
$$\begin{array}{r} 6\ 9\ 8 \\ \times\ 4\ 0 \\ \hline \end{array}$$

2. Solve these problems using rounded numbers.

a. A coloring book costs $6.85. How many could you buy with $30?	**b.** Calvin bought three bottles of juice for $3.47 each and two bags of pretzels for $5.83 each. About how much was the total cost?

3. Draw lines using a ruler.

a. 2 3/4 inches

b. 17 cm 6 mm

4. Figure out the pattern and continue it.

1   3   9   27   ____ ____ ____ ____ ____

5. Compare, and write < , >, or = in the boxes.

a. 100×70 ☐ 80×90 **b.** $175 + 5$ ☐ $240 - 60$ **c.** 9×45 ☐ 400

6. Solve. Write number sentence(s) on the empty lines to show your work.

Allison walks six miles a day, four days a week. How many miles does she walk in seven weeks?

_____

_____

Skills Review 45

1. Round these numbers to the nearest thousand, nearest ten thousand, and nearest hundred thousand.

number	645,498	462,784
to the nearest 1,000		
to the nearest 10,000		
to the nearest 100,000		

2. One foot is 12 inches. Convert between feet and inches.

a. 8 ft = _____ in

   12 ft = _____ in

b. 72 in = _____ ft

   90 in = _____ ft _____ in

3. Calculate in the right order.

$4728 + 7 \times (936 - 59)$

4. Solve the word problems. Also, write number sentences to show your calculations.

a. Vista Valley has a population of 472 people and Sunny City has a population that is six times that many. How many people live in Sunny City?

_____

_____

b. Madison has to work eight hours to earn $120. How much money does she earn in seven hours?
Hint: first find out how much money she earns in one hour.

Skills Review 46

1. Solve the pan balance equation.

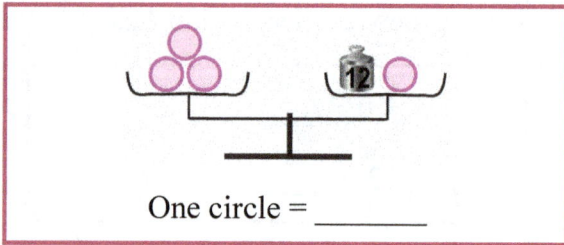

One circle = _____

2. Add in any order, and in parts.

a. $78 + 6 + 9 + 50 + 12 + 8$

b. $243 + 400 + 7 + 5 + 14 + 60$

3. Carolyn started cleaning the garage at 2:20 p.m. and finished three hours and 55 minutes later. What time did she finish?

4. Convert ounces to pounds and ounces.

a. 28 oz = ____ lb ____ oz	**b.** 48 oz = ____ lb ____ oz	**c.** 55 oz = ____ lb ____ oz
39 oz = ____ lb ____ oz	41 oz = ____ lb ____ oz	65 oz = ____ lb ____ oz

5. Abel gathered some twigs from his yard and recorded their lengths on the line plot below:

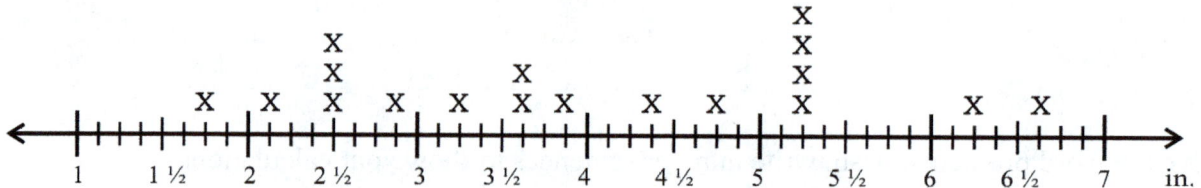

a. How many twigs are between 3 ½ and 7 inches long?

b. How long is the longest twig?

c. If Abel puts the three longest twigs end-to-end, how long is their combined length?

6. Divide.

a. $36 \div 4 =$ _____	**b.** $42 \div 6 =$ _____	**c.** $35 \div 7 =$ _____
d. $48 \div 8 =$ _____	**e.** $81 \div 9 =$ _____	**f.** $84 \div 12 =$ _____

7. Pam mailed two packages to her grandchildren. One package weighed 3 lb 12 oz and the other weighed 5 lb 14 oz.

a. What was their total weight in ounces?

b. In pounds and ounces?

Skills Review 47

1. Multiply — but first, estimate the result. Compare your final answer to your estimated answer. If there is a big difference, you might have an error somewhere.

a. 48×62	**b.** 91×33	**c.** 59×17
Estimate: _____ × _____	Estimate: _____ × _____	Estimate: _____ × _____
= _____	= _____	= _____

2. Skip-count backwards, subtracting 800 each time.

Start at 9,300

_____

_____

_____

_____

_____

3. One meter is 100 cm. Convert between meters and centimeters.

a. 9 m = _____ cm

14 m = _____ cm

b. 7 m 8 cm = _____ cm

12 m 30 cm = _____ cm

c. 900 cm = _____ m

548 cm = ____ m _____ cm

4. Write a number sentence with an unknown (x or ? or another symbol) for each problem. Then solve it.

a. Randy bought a backpack for $27.95 and a canteen for $17.60. After that, he had $287 left. How much money did he have originally?	
b. Erica had $106. Then she bought a game and now she has $81.04 left. How much did the game cost?	

Skills Review 48

1. Fill in the missing numbers so that both sides of the equals sign "=" have the same value.

| a. $8 \times 6 = 4 \times$ _____ | b. $6 \times 6 =$ _____ $\times 12$ | c. $6 \times$ _____ $= 9 \times 8$ |

2. Choose the right weight for each thing. Sometimes there are two possibilities.

a. a kitten	b. a car	c. a goat
8 oz 5 lb 4 oz	650 lb 2 T 11 T	10 oz 90 lb 150 lb

3. Draw a four-part rectangle to illustrate the multiplications. You don't have to draw to scale—a sketch is good enough.

$83 \times 65 =$

_____ $\times$ _____ $+$ _____ $\times$ _____

$+$ _____ $\times$ _____ $+$ _____ $\times$ _____

$=$

4. Write the numbers in expanded form.

a. 79,628

b. 342,415

Puzzle Corner Find the missing numbers.

a. $3 \times 49 + x = 189$	b. $x + 5 \times 18 = 127 + 4 \times 40$
$x =$ _____	$x =$ _____

Skills Review 49

1. Add or subtract.

a.	b.	c.
$\begin{array}{r} 72,409 \\ 36,158 \\ +\ 42,890 \\ \hline \end{array}$	$\begin{array}{r} 921,635 \\ 584,072 \\ +\quad 9,545 \\ \hline \end{array}$	$\begin{array}{r} 903,140 \\ -\ 746,358 \\ \hline \end{array}$

2. One kilogram is one thousand grams. Convert between kilograms and grams.

a. 7 kg = _____ g	b. 2 kg 300 g = _____ g	c. 6 kg 200 g = _____ g
4 kg = _____ g	9 kg 50 g = _____ g	3 kg 2 g = _____ g

3. Multiply.

a. $20 \times 50 =$ _____	b. $90 \times 700 =$ _____	c. $400 \times 400 =$ _____

4. Solve. Hint: in (b), first solve the scales on the left side.

a. One diamond weighs _____

b. One circle weighs _____

One triangle weighs _____

5. Write an addition or a subtraction with an unknown (*x* or ?). Solve it.

A book used to cost \$67.45; now the price is \$59.95. How much is the discount?

← original price →

_____

_____

Skills Review 50

1. Carolyn made a list of the ages of all the young children in her neighborhood:

4 6 2 5 10 7 1 4 2 6 5 6 3 2 4 3 6 9 5 8 7 4 10 9 6 10

a. Make a frequency table and a bar graph.

Age	Frequency

b. How many children are age 4 or older?

c. How many children are age 3 or younger?

2. Complete the *next* whole thousand.

a.	b.	c.
3,840 + _____ = _____	7,310 + _____ = _____	5,490 + _____ = _____

3. Estimate the results by rounding one or both factors. Don't round both numbers if you can multiply in your head just by rounding one factor.

a. 7×46	b. 23×49	c. 219×12
≈ ____ × _____ = _____	≈ _____ × _____ = _____	≈ _____ × _____ = _____

4. Complete. Remember: four cups = one quart, two pints = one quart, and four quarts = one gallon.

a.

Quarts	Cups
7	
9	

b.

Quarts	Pints
3	
8	

c.

Gallons	Quarts
1/2	
6	

Skills Review 51

1. Add the hours and minutes. Convert the sum of minutes into hours and minutes.

a.			
	9 h	45 m	
	6 h	20 m	
+	5 h	55 m	
=			

b.			
	2 h	38 m	
	4 h	57 m	
+	7 h	22 m	
=			

2. How much total time will it take to watch three nature documentaries that are each 1 hour and 35 minutes long?

3. Convert between liters and milliliters. Remember, there are 1,000 milliliters in one liter.

a.	b.	c.
8 L = _____ ml	12,000 ml = _____ L	6 L 95 ml = _____ ml
3 L = _____ ml	4,000 ml = _____ L	10 L 460 ml = _____ ml

4. Write the numbers and x into the bar model. Then solve for x.

a.

$$94 - x = 36$$

$x =$ _____ $-$ _____ $=$ _____

b.

$$72 - 13 = x$$

$x =$

5. Solve.

a. $500 \times$ _____ $= 40,000$	b. $760 \times$ _____ $= 760,000$	c. _____ $\times 400 = 240,000$
_____ $\times 20 = 600$	_____ $\times 100 = 43,000$	$700 \times 30 =$ _____

6. One yard of ribbon costs $1.60. Fill in the table.

Dollars	$1.60									
Yards	1	2	3	4	5	6	7	8	9	10

Skills Review 52

1. Find what number x stands for.

a. $63 \div x = 9$	b. $42 \div x = 7$	c. $x \div 4 = 8$	d. $x \div 8 = 6$
$x =$ _____	$x =$ _____	$x =$ _____	$x =$ _____

2. The line graph shows how much money Nancy had in her savings over a seven-month period.

   a. During which month did her savings increase the most?

   b. How much more money did she have in July than in January?

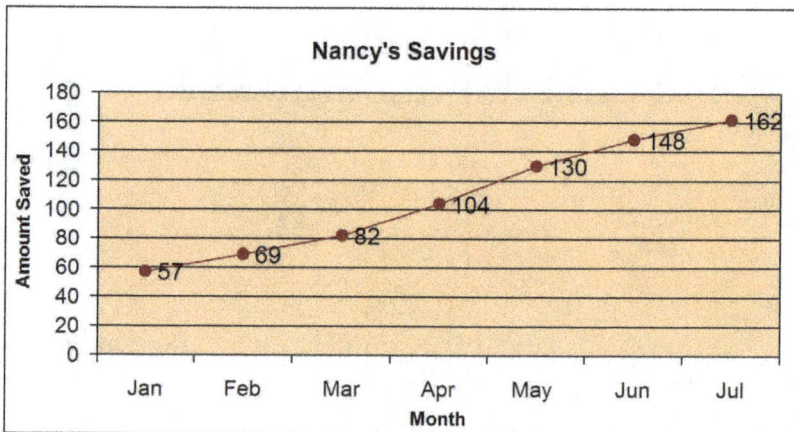

Nancy's Savings

   c. Nancy wants to buy a painting that costs $249. How much more money does she need to save?

3. Subtract with money amounts. Check by adding!

a.	b.
$\begin{array}{r} \$ 6 0 . 0 0 \\ - \quad 4 9 . 0 8 \end{array}$ +	$\begin{array}{r} \$ 5 0 0 . 0 0 \\ - \quad 3 6 0 . 2 0 \end{array}$ +

4. Write < or > between the numbers.

a. 87,300 87,030	b. 92,005 92,050	c. 30,310 303,100
d. 450,231 405,231	e. 6,920 6,902	f. 2,720 27,200

5. Solve mentally.

$(550 - 50) \times 400 =$ _____

$900 + 3 \times 700 =$ _____

6. Multiply.

a.
$\begin{array}{r} 9\ 4 \\ \times\ 9\ 0 \\ \hline \end{array}$

b.
$\begin{array}{r} 6\ 8 \\ \times\ 5\ 0 \\ \hline \end{array}$

c.
$\begin{array}{r} 3\ 0\ 7 \\ \times\ 7\ 0 \\ \hline \end{array}$

Skills Review 53

1. First, fill in the top row, continuing the pattern it has. Then solve what number is being subtracted and complete the bottom row.

n	38	128	218	308					
$n -$ ____		99							

2. Fill in the missing numbers. Write the area of the *whole* rectangle as a SUM of the areas of the *smaller* rectangles. Also find the total area.

9×147

= ____ $\times$ _____ + ____ $\times$ ____ + ____ $\times$ ____

=

3. How much time passes? Do it in two parts, since one time has a.m. and the other has p.m.

a. From 9:46 p.m. till 4:28 a.m.	**b.** From 6 a.m. till 7:32 p.m.

4. Measure the length of the items named below using a ruler that has marks for eighth parts of an inch. A sewing tape measure may also be used. You choose two items to measure.

Item	Length/width	Item	Length/width
an envelope			
your index finger			

5. Divide. Mark off the problem if it is impossible to do.

a. $0 \div 9 =$ ____	**b.** $60 \div 12 =$ ____	**c.** $36 \div 6 =$ ____	**d.** $0 \div 0 =$ ____
$56 \div 7 =$ ____	$48 \div 0 =$ ____	$81 \div 9 =$ ____	$96 \div 8 =$ ____

1. What is missing?

   a. $7{,}248 = 7{,}000 + 200 +$ _____ $+ 8$ **b.** $9{,}561 = 500 + 60 +$ _____ $+$ _____

2. Use rounded numbers to solve the problem.

   > *Round the numbers to the nearest thousand.*
   >
   > There were 73,258 visitors at the state fair on Friday and 117,721 visitors on Saturday.
   >
   > This means there were about _____ visitors on Friday, and about _____ visitors
   >
   > on Saturday. The two days had approximately _____ visitors in all.
   >
   > There were about _____ more visitors on Saturday than on Friday.

3. **a.** Measure all the sides of this figure to the nearest eighth of an inch. Write the measurements next to the sides of the figure.

   b. Measure its sides also in centimeters and millimeters.

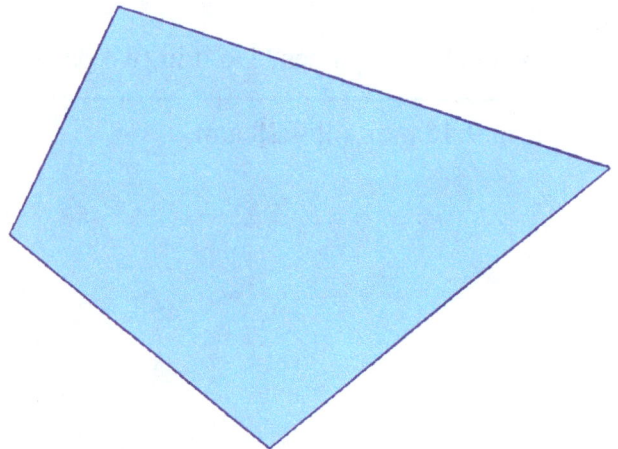

4. Divide.

a. $300 \div 6 \ =$ _____	**b.** $5{,}400 \div 9 =$ _____	**c.** $7{,}200 \div 800 =$ _____
$300 \div 60 =$ _____	$5{,}400 \div 90 =$ _____	$7{,}200 \div 8 \ =$ _____

5. Solve the problem.

   > Bethany earns $240 a week and saves half of that each week.
   > She wants to buy a laptop that costs $1,800 but she has only
   > saved $960. How many more weeks will she need to save?

Skills Review 55

1. Are these situations reasonable? Answer true or false.

    a. Ouch! I burned my tongue! This tea must be at least 40° Fahrenheit! _____

    b. What a lovely summer day! The temperature outside is 28° Celsius. _____

    c. I don't have a fever. My temperature is 39° Celsius. _____

    d. The temperature outside is 72° Fahrenheit. Let's go skiing! _____

    e. My pot of soup won't boil until it reaches 212° Fahrenheit. _____

2. Solve.

a. $12 \times (300 \div 30) =$ _____	**b.** $(90 + 70) \div 40 =$ _____

3. A group of 9 friends decided to go on vacation together.
   Each one paid $798 for a round-trip airplane ticket.

    a. Estimate how much money they spent in total.

    b. Now calculate the exact amount.

4. Draw a bar model and fill it in. Then solve and write a number sentence.

Carolyn had $211. She bought some towels and some soap, and had $182 left. If the towels cost $27, how much did the soap cost?

5. Find the starting time.

    a. From _____ : _____ p.m. till 11:18 p.m. is 1 hour 20 minutes.

    b. From _____ : _____ a.m. till 6:42 a.m. is 2 hours 40 minutes.

6. Find half and the double of the given numbers.

Half the number			
Number	94	350	726
Double the number			

7. Subtract mentally.

a. $80 - 36 =$ _____
b. $200 - 79 =$ _____
c. $53 - 17 =$ _____

Skills Review 56

1. Fill in the table.

×	5	0	7	3	12	4	10	2	9	1	6	11	8
9													
7													
8													
6													
5													

2. Change the 24-hour clock times to a.m. / p.m. times.

a. 16:48	b. 13:13
_____ : _____	_____ : _____
c. 22:29	d. 19:09
_____ : _____	_____ : _____

3. Solve.

a. $38 \div 9 = $ _____ R _____

   $52 \div 6 = $ _____ R _____

b. $87 \div 12 = $ _____ R _____

   $90 \div 8 = $ _____ R _____

4. Solve.

a. Brenda is 5 ft 7 inches tall, and Kylie is 4 ft 9 inches tall. How many inches taller is Brenda than Kylie?

b. Sammy was 3 feet 8 inches tall on his sixth birthday, and 4 feet 2 inches tall on his eighth birthday. How much did he grow in those two years?

If he keeps growing at the same rate, how tall will he be when he is 10?

When he is 12?

5. Add or subtract.

a.
```
  7 0 9 , 2 6 8
+   3 4 , 9 5 0
_____
```

b.
```
  3 2 0 , 6 0 0
−   6 1 , 7 5 2
_____
```

c.
```
  8 1 5 , 0 0 0
−     8 , 7 4 9
_____
```

Skills Review 57

1. What are these numbers?

   a. The tens and ones digits are divisible by four, the ones digit being the larger of the two. The digits in the ten thousands and hundred thousands places are both odd, the latter being two more than the former, and their sum is 12. The rest of the digits are zeros.

   b. The hundreds digit is double the tens digit, the thousands digit is double the hundreds digit, the ten thousands digit is double the thousands digit, and the ones digit is zero.

2. Multiply.

a.	**b.**	**c.**	**d.**
$\begin{array}{r} 79 \\ \times\ 46 \\ \hline \end{array}$	$\begin{array}{r} 25 \\ \times\ 38 \\ \hline \end{array}$	$\begin{array}{r} 92 \\ \times\ 63 \\ \hline \end{array}$	$\begin{array}{r} 40 \\ \times\ 57 \\ \hline \end{array}$

3. Brenna went to bed at 9:30 pm and slept for six hours and 50 minutes. What time did she wake up?

4. Find the correct number to replace x.

a. $800,000 - x = 340,000$	**b.** $x + 70,000 = 220,000$	**c.** $x + x = 90,000$
$x = $ _____	$x = $ _____	$x = $ _____

5. Solve the problems.

   a. Greg rides his bike a distance of 2 km 500 m to work and the same distance back home. He works five days a week. What distance does he ride in 13 weeks?

   b. One side of a rectangle measures 8 cm 6 mm and another side measures 5 cm 9 mm. What is its perimeter?

Skills Review 58

1. Convert pounds and ounces to ounces. Remember, one pound is 16 ounces.

a. 7 lb = _____ oz	**b.** 2 lb 9 oz = _____ oz	**c.** 8 lb 5 oz = _____ oz
4 lb = _____ oz	6 lb 3 oz = _____ oz	3 lb 15 oz = _____ oz

2. Kevin asked his classmates about their favorite school subject, and made a list of their answers:

P.E. math science P.E. art history English math art art science P.E. math math history P.E. art math English science history P.E. P.E. science art

a. Make a frequency table and a bar graph.

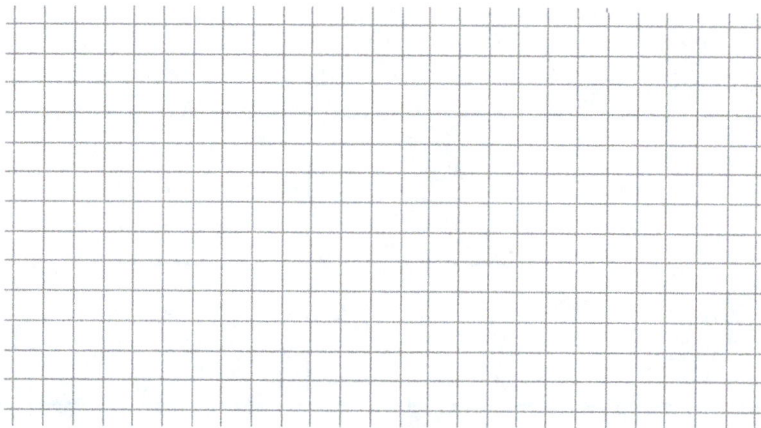

Subject	Frequency

b. What was the most popular school subject?

c. How many classmates does Kevin have?

3. Divide. Mark out the problem if it is impossible to do.

a. $0 \div 9 =$ _____
b. $22 \div 0 =$ _____
c. $64 \div 8 =$ _____

4. Multiply.

a. $20 \times 700 =$ _____	**d.** $800 \times$ _____ $= 8{,}800$
b. $6 \times 50{,}000 =$ _____	**e.** $30 \times$ _____ $= 27{,}000$
c. $400 \times 12 =$ _____	**f.** _____ $\times 6{,}000 = 42{,}000$

5. Divide and find the remainder by subtracting!

a. $7 \overline{)5\ 9}$

b. $4 \overline{)3\ 5}$

c. $6 \overline{)6\ 8}$

d. $8 \overline{)7\ 6}$

Skills Review 59

1. Find the numbers that go on the empty lines.

a. $4 \times 8 - 6 = 5 \times$ ___ $+ 1$	**b.** $8 \times 20 = 30 +$ _____	**c.** $26 + 24 = 3 \times$ ___ $+ 2$

2. Mary made a total of 15 cups of coffee for a business meeting. How many ounces of coffee is that?

3. Write how many of each bill you need to make change. Use mental math.

Item cost	Money given	Change needed	$50 bill	$20 bill	$5 bill	$1 bill
a. $64	$100					
b. $28	$50					

4. Divide. Check each one by multiplying.

a. $738 \div 3$ Check:	**b.** $568 \div 8$ Check:

5. Circle the number sentence that fits the problem. Then solve for x.

a. Kim bought a dress for $16 and now she has $29 left. $29 - x = \$16$ OR $x - \$16 = 29$ $x =$ _____	**b.** Carl had 38 baseball cards. Then, he bought some more and now he has 54. $38 + x = 54$ OR $38 + 54 = x$ $x =$ _____

Skills Review 60

1. Fill in.

a. Kay baked 60 cookies. She gave 12 to Mr. Hill, which was one-_____ part of them.

Division sentence: _____ ÷ _____ = _____

b. Caleb gave one-third (eight) of his toy cars to Jeremy. So originally, he had _____ toy cars.

Division sentence: _____ ÷ _____ = _____

2. Convert these times.

a. 7 h = _____ min	**b.** 10 h 17 min = _____ min	**c.** 180 min = _____ h
3 h = _____ min	12 h 47 min = _____ min	420 min = _____ h

3. Round these numbers to the nearest dollar and to the nearest ten dollars.

n	$44.62	$17.29
rounded to nearest dollar		
rounded to nearest ten dollars		

4. Circle the heaviest amount.

a. 7 kg 60 g   OR   7,006 g   OR   780 g

b. 9 kg 800 g   OR   9,080 g   OR   90,800 g

c. 5 kg 3 g   OR   503 g   OR   530 g

5. First, fill in the top row, continuing the pattern it has. Then fill in the bottom row, keeping in mind that each number in the bottom row is 89 less than the number in the top row.

n	470	520	570					
$n - 89$								

6. Solve. Write number sentence(s) on the empty lines to show your work.

The Johnson Family eats 70 eggs a week for breakfast, and Mrs. Johnson uses an average of 8 eggs a week for baking. Based on this, how many eggs would they use in five weeks?

_____

_____

Skills Review 61

1. Add.

a. 4 L 725 ml 350 ml + 3 L 240 ml =	b. 6 L 500 ml + 2,750 ml	c. 840 ml + 320 ml + 5 L 30 ml

2. Divide. Use the grids below. Check each one by multiplication.

a. 3798 ÷ 9   Check:

b. 2352 ÷ 7   Check:

3. It takes Marsha 25 minutes to drive to the store from her house.
   She left for the store at 2:45 and needs to be back home by 4:30.

   a. What time should she start driving home?

   b. How much time does she have to shop?

4. Al bought 30 packages of construction paper with 60 sheets
   in each package. Each package had 10 different colors in equal amounts.
   How many sheets of yellow paper were there in total?

5. Subtract mentally.

a. 210 − 170 = _____	b. 1,000 − 430 = _____	c. 700 − 29 = _____

Skills Review 62

1. The graph shows the snowfall in the first two weeks of 2017 at Steamboat Resort in Colorado.

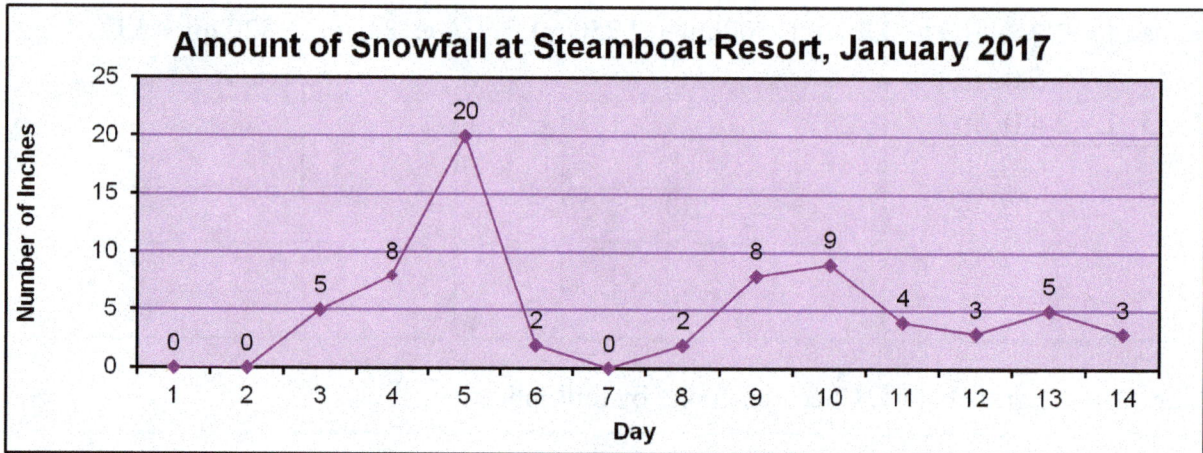

Amount of Snowfall at Steamboat Resort, January 2017

(graph: Number of Inches vs. Day — Day 1: 0, Day 2: 0, Day 3: 5, Day 4: 8, Day 5: 20, Day 6: 2, Day 7: 0, Day 8: 2, Day 9: 8, Day 10: 9, Day 11: 4, Day 12: 3, Day 13: 5, Day 14: 3)

   a. Which week was "snowier" — the week of January 1-7, or the following week?

   b. The average daily snowfall for the second week was 4 7/8 inches.
   What was the average daily snowfall for the first week of January?

2. Round the numbers as the dashed line indicates (to the underlined digit).

a. 76̲5,472 ≈	**b.** 328̲,910 ≈	**c.** 87,3̲51 ≈

3. Elsa baked 87 cookies and packaged them in packages
   of six. How many full packages did she have?

4. There will be 99 people attending the school picnic.
   If the teachers plan on seating eight people at each
   table, how many tables will they need?

5. Draw a four-part rectangle to illustrate the multiplication. You don't have to draw to scale —
   a sketch is good enough.

   $46 \times 29 =$

   _____ × _____ + _____ × _____

   + _____ × _____ + _____ × _____

   =

Skills Review 63

1. Multiply.

a. $\begin{array}{r} 79 \\ \times\ 35 \\ \hline \end{array}$

b. $\begin{array}{r} 82 \\ \times\ 40 \\ \hline \end{array}$

c. $\begin{array}{r} 963 \\ \times\ 80 \\ \hline \end{array}$

d. $\begin{array}{r} 54 \\ \times\ 66 \\ \hline \end{array}$

e. $\begin{array}{r} 218 \\ \times\ 90 \\ \hline \end{array}$

2. Solve the problems.

a. If a gallon of ice cream costs $9.84, how much would one pint cost?

b. Joy bought six shirts for $50.70. How much did one cost?

3. Solve.

a. $70 + (124 - 40) \div 12 =$ _____

b. $30 \times 40 \div 60 =$ _____

c. $200 \div 20 \times 90 - 80 =$ _____

4. Add or subtract, thinking in whole thousands.

a. $580,000 + 60,000 =$

b. $300,000 - 40,000 =$

c. $720,000 - 80,000 =$

d. $660,000 + 90,000 =$

5. Draw lines using a ruler.

a. 6 3/8 inches

b. 97 mm

Skills Review 64

1. Calculate.

a. $\dfrac{2}{5}$ of 30 is _____.	**b.** $\dfrac{2}{7}$ of 42 is _____.	**c.** $\dfrac{2}{9}$ of 63 is _____.
$\dfrac{3}{5}$ of 30 is _____.	$\dfrac{5}{7}$ of 42 is _____.	$\dfrac{6}{9}$ of 63 is _____.
$\dfrac{4}{5}$ of 30 is _____.	$\dfrac{6}{7}$ of 42 is _____.	$\dfrac{8}{9}$ of 63 is _____.

2. **a.** Measure all the sides of this figure to the nearest eighth of an inch. Write the measurements next to the sides of the figure.

   b. Also, measure its sides in millimeters.

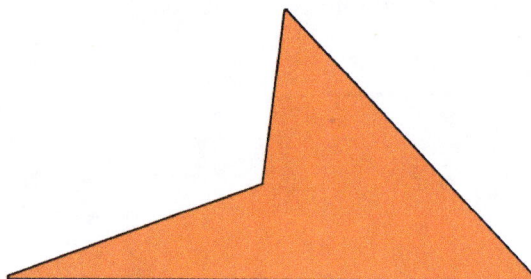

3. Estimate. Round *both* the dividend and the divisor to the nearest ten.

a. 503 ÷ 45	**b.** 752 ÷ 27	**c.** 396 ÷ 82
≈ _____ ÷ _____	≈ _____ ÷ _____	≈ _____ ÷ _____
= _____	= _____	= _____

4. It is time to test your knowledge with missing factor problems!

a. _____ × 8 = 56	**b.** _____ × 5 = 40	**c.** _____ × 9 = 63	**d.** _____ × 4 = 28
_____ × 3 = 21	_____ × 11 = 132	_____ × 7 = 49	_____ × 6 = 54

5. Of Golden Valley's 3,268 residents, 1,018 are men and 1,294 are women. How many are children?

6. Remember, one yard is 3 feet. Convert between yards and feet.

a. 7 yd = _____ ft	**b.** 9 yd 2 ft = _____ ft	**c.** 39 ft = _____ yd
18 yd = _____ ft	12 yd 1 ft = _____ ft	48 ft = _____ yd

Skills Review 65

1. Estimate the cost. Round one or both numbers so that you can multiply in your head.

a. 56 bicycles at $120 each	**b.** 718 combs at 67¢ each
≈	≈

2. Write the numbers in expanded form.

   a. 74,295

   b. 306,187

3. The temperature rises or falls. Write the new temperature.

Now	temperature rises 4° C	**After**	**Now**	temperature falls 5° C	**After**	**Now**	temperature falls 5° C	**After**
a. -6° C		_____	**b.** -8° C		_____	**c.** -2° C		_____

4. Calculate. Give your answer using whole kilometers and meters. Remember, one kilometer is 1,000 meters.

   a. 7 km 600 m + 4 km 800 m

   b. 12 km 900 m + 9 km 700 m

5. Add or subtract.

a.	**b.**	**c.**
$\begin{array}{r} 739,087 \\ +\ \ 82,624 \\ \hline \end{array}$	$\begin{array}{r} 524,306 \\ -\ \ 37,459 \\ \hline \end{array}$	$\begin{array}{r} 406,700 \\ 860,000 \\ +374,932 \\ \hline \end{array}$

6. The owner of a company invited his 270 employees to a dinner, but 2/9 of them were unable to attend. How many employees attended the dinner?

Skills Review 66

1. Find the missing factors.

a. $36 \times \underline{\hspace{1cm}} = 3600$
b. $71 \times \underline{\hspace{1cm}} = 710$
c. $\underline{\hspace{1cm}} \times 53 = 53{,}000$

2. Complete the table so that each row or column of numbers adds up to 50,000.

12,600		7,400	11,100	= 50,000
4,500	9,700		14,100	= 50,000
	6,300	8,700		= 50,000
10,100	15,100	12,200		
= 50,000	= 50,000	= 50,000		

3. Choose the right weight for each thing. Sometimes there are two possibilities.

a. a pillow	**b.** a watermelon	**c.** a man
4 oz 4 lb 40 lb	20 lb 2 oz 500 lb	150 lb 1 T 240 lb

4. Rhonda bought three coloring books for $6.75 each and a set of coloring pencils for $12.89, and had $13.64 left. How much money did Rhonda have initially?

My estimate:

My calculations:

5. Calculate.

a. $630{,}547 + 98{,}593$

b. $72{,}305 - 4{,}068$

All three of us are in the table of 11! If you take six away from any of us, you will get a number that is divisible by 2 but not divisible by 4.

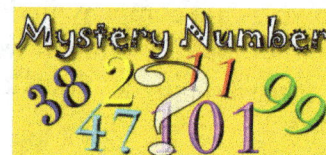

Mystery Number
38 26 31 11 99
47 101

Skills Review 67

1. Solve.

One circle weighs _____

One square weighs _____

2. Fill in.

Days	Hours
5	
3	
6	
4	

Years	Months
7	
2	
5	
3	

3. Check if each number is prime or composite. If it is composite, write it as a multiplication.

a. 83 is prime/composite If a composite: 83 = ____ × ____	**b.** 68 is prime/composite If a composite: 68 = ____ × ____	**c.** 57 is prime/composite If a composite: 57 = ____ × ____

4. Alex bought two gallons of orange juice, Carol bought 10 pints, and Janice bought seven quarts. How many *gallons* of juice did they buy in total?

5. Divide. Use the grids below. Check each one by multiplication.

a. 1504 ÷ 8   Check:

b. 2364 ÷ 6   Check:

Skills Review 68

1. Write the numbers in order.

    3 1 0 0 6 3   1 0 0 3 6   3 1 0 6 3   1 0 6 3 0   3 0 1 0 6 0   1 6 3 0 3

    _____ < _____ < _____ < _____ < _____ < _____

2. Write an equation (a multiplication with an unknown) for the area. Then solve.

    The area of a rectangle is 84 m^2,
    and its one side measures 7 m.
    How long is the other side (s)?

3. Find all the factors of the given numbers. (Hint: check each number from 1 through 10 if it is a factor.)

a. 72	**b.** 54
Check 1 2 3 4 5 6 7 8 9 10	Check 1 2 3 4 5 6 7 8 9 10
factors: _____	factors: _____

4. Describe a situation to fit these temperatures.

    a. 97°F

    b. 25°F

5. Write a missing addend problem that matches the bar model. Then solve it by subtracting.

    Eve has 463 beads. Some are red, and
    294 are blue. How many beads are red?

    ← total ____ →

    _____ + _____ = _____

    $x =$ _____ − _____ = _____

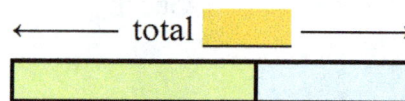

6. Four cans of soup cost $13.40.
   How much would three cans cost?

7. Make up a division with...

a. ...a quotient of 1.
_____ ÷ _____ = _____
b. ...a dividend of 0.
_____ ÷ _____ = _____

Skills Review 69

1. **a.** A square has an area of 36 cm². How long is its side?

   b. What is its perimeter?

2. Find the starting time.

a. From _____ : _____ p.m. till 3:00 p.m. is 50 minutes.
b. From _____ : _____ p.m. till 9:43 p.m. is 25 minutes.
c. From _____ : _____ a.m. till 1:15 p.m. is 2 hours 40 minutes.

3. Mr. Jacobs bought 30 sheep at $500 each.
   What was the total cost?

4. In March 2016, Key West had 68,894 visitors who arrived on cruise ships. In March 2017, there were 77,623. How many more people went on a cruise to Key West in March 2017 than in March 2016?

5. Convert between kilograms and grams.

a. 7 kg = _____ g	**b.** 9,000 g = _____ kg	**c.** 4 kg 300 g = _____ g

6. The bar graph shows how many e-mails Pamela received during a specific 5-day workweek.

   a. Find the *average* number of e-mails that Pamela received daily.

   Pamela's E-mails

Day	How many E-mails
Friday	53
Thursday	71
Wednesday	42
Tuesday	68
Monday	56

   b. The following workweek was less busy, and Pamela received just 3/5 as many emails as she received the previous week. How many e-mails was that?

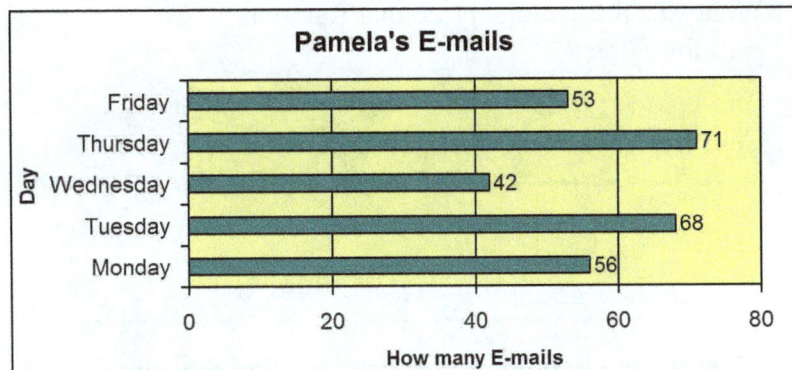

Skills Review 70

1. Write a number sentence with an unknown (*x* or *?* or another symbol) for the problem. Then solve it.

Bill bought four shirts for $7 each and now he has $33 left. How much money did he have originally?	

2. Fill in the missing numbers. Write the area of the *whole* rectangle as a SUM of the areas of the *smaller* rectangles. Also find the total area.

$49 \times 85 =$ _____ $\times$ _____ $+$ _____ $\times$ _____

$+$ _____ $\times$ _____ $+$ _____ $\times$ _____

$=$

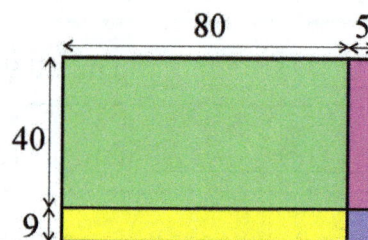

3. Three children baked cookies and packaged them in different-sized packages. Complete the chart.

Child	Marsha	Kyle	Bonnie
Number of cookies baked	76	87	91
How many in each package	8	12	6
Number of full packages			
How many cookies left			

4. Ramona went shopping and bought four new dresses. She paid $14, $18, $21, and $19. What was the average price that Ramona paid for a dress?

5. Convert between liters and milliliters.

a.	b.	c.
8 L = _____ ml	3,430 ml = ____ L _____ ml	9,000 ml = ____ L

6. Write if each figure is a line, ray, or a line segment.

a. _____   b. _____   c. _____

Skills Review 71

1. Multiply.

   a.   7 8   **b.**   7 3   **c.**   6 0 5   **d.**   9 1 3   **e.**   9 9

       × 9 0   × 5 6   × 8 0   × 7 0   × 8 1

2. Solve.

a. $8 \times (30 + 40) \div 20 =$ _____	**b.** $460 - 20 \times 20 \div 8 =$ _____

3. Measure the lines to the nearest fourth of an inch. Also measure them in centimeters and millimeters.

       a. _____ in. or _____ cm _____ mm

       b. _____ in. or _____ cm _____ mm

4. Find the pattern and continue it.

  2,130   2,090   2,050   _____ _____ _____ _____ _____

5. Scott picked 186 apples and divided *half* of them as evenly as possible between four of his neighbors. How many apples did each neighbor get?

6. Use rounded numbers to solve the problem.

> *Round the numbers to the nearest thousand.*
>
> There are 267,418 males and 391,710 females living in Cinnamon City.
>
> This means there are about _____ males and about _____ females.
>
> There are approximately _____ people in all.
>
> There are about _____ more females than males in Cinnamon City.

1. Find the missing numbers. The sum of any two adjacent (side-by-side) numbers is the number directly above them.

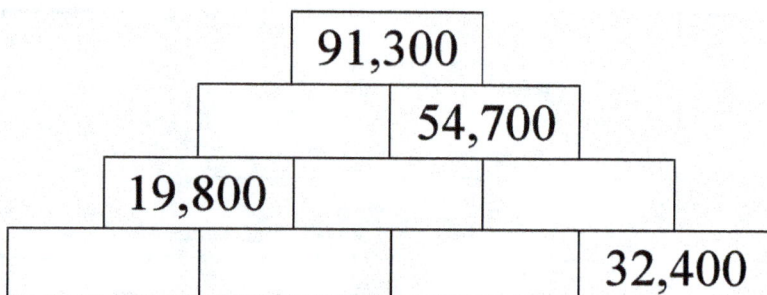

```
              91,300
                    54,700
        19,800
                        32,400
```

2. Calculate.

a. $\dfrac{3}{4}$ of 48 is _____.	
b. $\dfrac{2}{6}$ of 54 is _____.	
c. $\dfrac{7}{10}$ of 700 is _____.	
d. $\dfrac{5}{8}$ of 160 is _____.	

3. Molly made a line graph to show how much weight her little sister gained each year from age one to age 10.

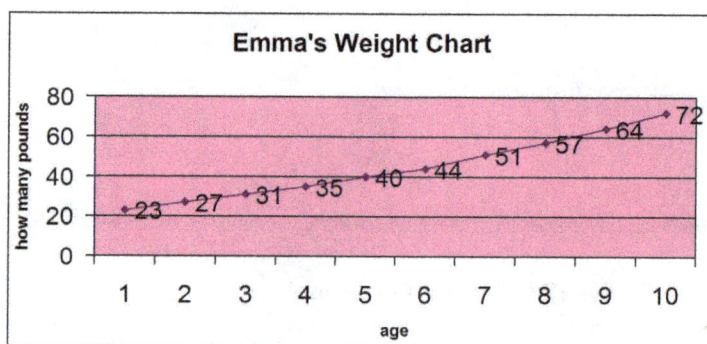

   a. Between which two consecutive ages did Emma gain the most weight?

   b. How much more did Emma weigh when she was seven than when she was three?

   c. How much weight did Emma gain in total from age one to age 10?

Emma's Weight Chart

how many pounds: 80, 60, 40, 20, 0

23 27 31 35 40 44 51 57 64 72

age: 1 2 3 4 5 6 7 8 9 10

4. Measure this angle using your protractor.

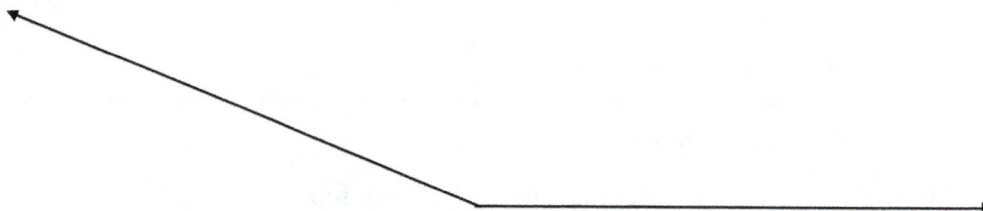

5. Convert between centimeters and millimeters.

a. 6 cm = _____ mm	**b.** 3 cm 9 mm = _____ mm	**c.** 68 mm = _____ cm _____ mm

Skills Review 73

1. Draw angles for these measurements.

a. 43°	**b.** 128°

2. Jason bought a laptop for $528.84 and a chair for $142.27. Estimate the total cost.

3. Each of Mrs. Carson's 39 students brought 24 clothespins to school for a craft project. Mrs. Carson estimated that the total number of clothespins would be $40 \times 25 = 1,000$. How many fewer clothespins than 1,000 do they have?

4. Find what number the unknown stands for.

a. $y \div 7 = 900$	**b.** $s \div 9 = 60$	**c.** $4,200 \div w = 700$
$y =$ _____	$s =$ _____	$w =$ _____

5. Oh no! Little Jimmy pulled a bunch of carrots in the garden, and most of them were not ready to harvest yet! Big sister Lucy decided to measure the carrots and make a line plot.

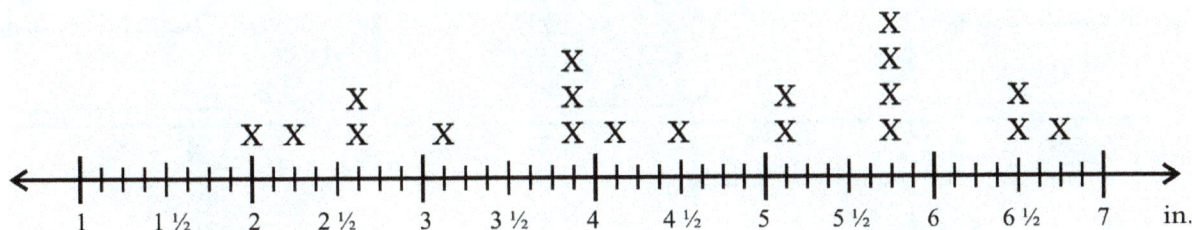

```
                                              X
                              X               X
                  X           X       X       X       X
              X X X   X       X X X   X       X       X X
        ←—┼┼┼┼┼┼┼┼┼┼┼┼┼┼┼┼┼┼┼┼┼┼┼┼┼┼┼┼┼┼┼┼┼┼┼┼┼┼┼┼—→
           1   1½   2   2½   3   3½   4   4½   5   5½   6   6½   7   in.
```

   a. Finish drawing the line plot by adding the x-marks for the carrots that measured 1 6/8 inches, 3 5/8 inches, 5 3/8 inches, and 6 2/8 inches.

   b. How many carrots were between two inches and five inches long?

Skills Review 74

1. Solve. (Remember, one mile is 5,280 feet.)

> Sarah's house is 3,960 ft away from her grandma's house. One day, Sarah walked to her grandma's house and back five times. *About* how many whole miles did she walk?
>
> _____

2. Starting at the top, find two different ways through the maze by coloring the number that is **one-fourth** of the previous number.

128,000	240,000	164,000	320,000
61,000	32,000	80,000	82,000
8,000	20,000	15,000	25,000
5,000	2,000	7,000	4,000
900	1,250	500	1,500

3. Fill in the table.

×	3	6	9	12
0				
2				
4				
6				
8				

4. Carmen and Erica bought two games for $27.60 each and a volleyball for $14.50. They shared the cost equally. How much did each one pay?

5. Estimate the angle measures in degrees.

a. _____   b. _____   c. _____

6. Divide and find the remainder by subtracting.

a. $6\overline{)59}$ b. $9\overline{)78}$ c. $4\overline{)51}$ d. $7\overline{)66}$

Skills Review 75

1. Estimate the results by rounding one or both factors. Don't round both numbers if you can multiply in your head just by rounding one factor.

a. 7×88	**b.** 12×63	**c.** 218×17
$\approx$ _____ $\times$ _____ = _____	$\approx$ _____ $\times$ _____ = _____	$\approx$ _____ $\times$ _____ = _____

2. Convert pounds and ounces to ounces.

a. 7 lb = _____ oz	**b.** 6 lb 12 oz = _____ oz	**c.** 3 lb 15 oz = _____ oz
4 lb = _____ oz	9 lb 3 oz = _____ oz	8 lb 9 oz = _____ oz

3. Erin can text on her cell phone at a speed of 42 wpm (words per minute), whereas the average texting speed for people with the same type of cell phone has been found to be only 30 wpm. In 3 minutes, how many more words can Erin text than the average person?

4. Find rays, lines, and line segments that are either parallel or perpendicular to each other. Use the symbols ⊥ and ∥.

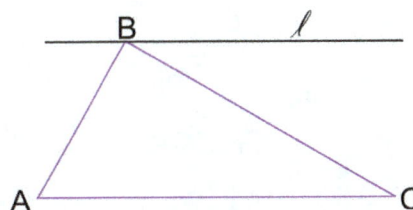

5. A building has two rooms. One room is 22 ft by 18 ft, and the other room is 22 ft by 35 ft.
   a. Draw a sketch of the building.

   b. Find its area.

Puzzle Corner Find the missing numbers.

a. $2{,}000 - 63 - 2 \times$ _____ $- 90 = 1{,}807$ **b.** $5{,}000 - 10 \times$ _____ $- 35 - 390 = 4{,}495$

Skills Review 76

1. Change the times to the 24-hour clock times.

a. 9:40 a.m.	**b.** 6:00 p.m.	**c.** 10:20 p.m.	**d.** 4:02 a.m.
_____ : _____	_____ : _____	_____ : _____	_____ : _____

2. A cheesecake that weighs 480 g is cut into eight equal-sized pieces.

   a. Annie eats 3/8 of it. How much do those slices weigh?

   b. Peter eats 420 grams of another same kind of cheesecake.
   How many slices is that?

3. Measure all the angles in the figures. Then add their angle measures. Verify that you get 360°.

a.

b.

4. Fill in. Draw a rectangle for the problem.

Sides 6 and _____ units;
area 30 square units,
perimeter _____ units.

5. Write the temperature each thermometer shows.

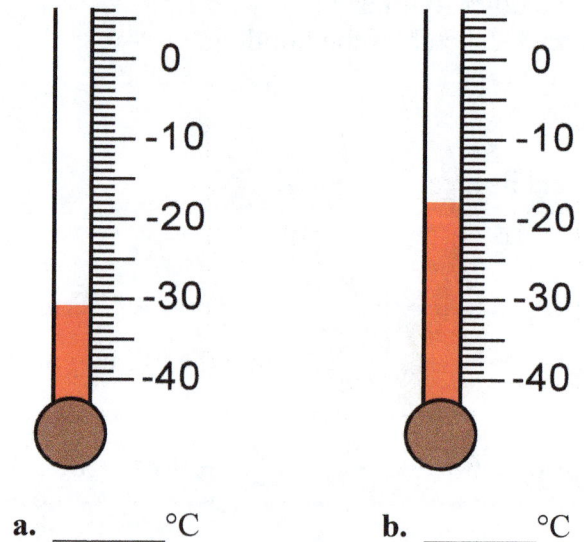

a. _____°C

b. _____°C

Skills Review 77

1. Answer the questions. You may need long division.

a. Is 728 divisible by 9?	**b.** Is 426 divisible by 6?

2. Fill in.

Minutes	Seconds
2	
7	
4	
9	
3	
8	
5	
10	

3. Label each quadrilateral as a rhombus, a trapezoid, or a parallelogram.

a.	b.	c.	d.

4. Jenna bought 2 kg of tomatoes, 750 g of peppers, 340 g
of onions, and 830 g of cucumbers. How much did the
vegetables weigh in total?

5. Add or subtract.

a.
$$875{,}063$$
$$+\ 68{,}947$$

b.
$$320{,}415$$
$$-\ 83{,}726$$

c.
$$446{,}379$$
$$+\ \ \ 6{,}805$$

1. Color the prime numbers green and
   the composite numbers blue.

165	95	157	117	74
74	151	67	29	169
17	89	191	13	139
173	132	109	63	101
27	128	103	81	121

2. Convert.

a. 2 pt = _____ C

b. 7 qt = _____ pt

c. 5 gal = _____ qt

d. 3 qt = _____ C

3. Label the triangles in the pictures as right, acute, or obtuse.

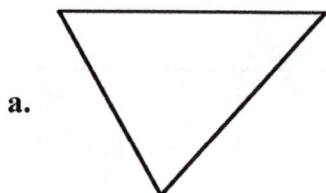

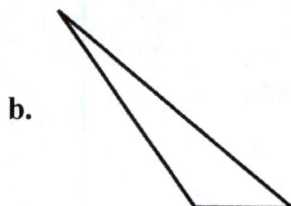

a.

b.

c.

4. How much time passes? Do it in two parts,
   since one time has a.m. and the other has p.m.

a. From 6:38 p.m. till 7:42 a.m.

b. From 3 a.m. till 8:14 p.m.

5. Multiply.

a. $500 \times 500 =$ _____

b. $40 \times 90 =$ _____

6. Color the thermometer to show the temperature
   on the thermometers. Notice the scale carefully:
   it is not by ones.

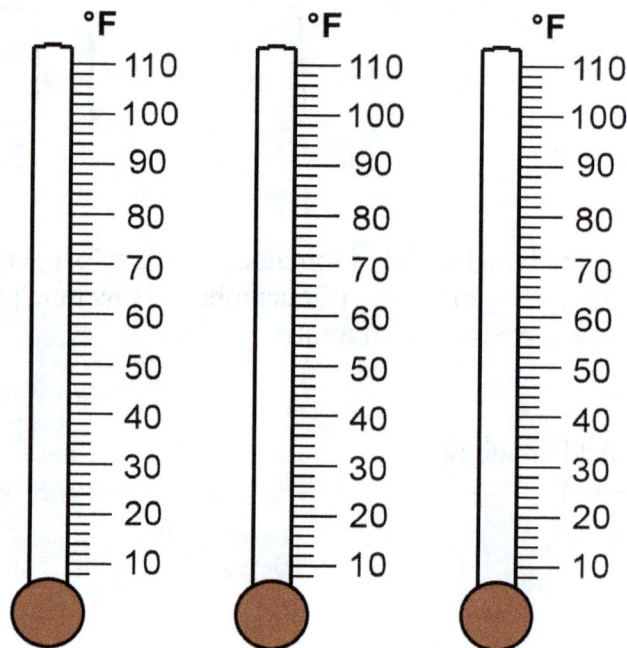

°F   °F   °F
110   110   110
100   100   100
90   90   90
80   80   80
70   70   70
60   60   60
50   50   50
40   40   40
30   30   30
20   20   20
10   10   10

a. 24°F   **b.** 86°F   **c.** 42°F

Skills Review 79

1. On a blank sheet of paper, draw these angles without using a protractor. In other words, estimate. After you have drawn the angle, measure it with a protractor to see how close you were.

   a. 75°   **b.** 135°

2. Write < or > between the numbers.

a. 5 6 2 0	5 6 2 0 0
b. 8 3 0 5 1 2	8 0 3 5 1 2
c. 6 2 3 0	6 0 3 2

3. Color all the numbers that are factors of the number in the middle.

   2 10 5 4 8 9 **63** 3 6 7

   2 6 9 5 4 8 **42** 7 3 10

   9 6 4 2 8 7 **72** 3 5 10

4. Clarissa made fruit punch for a party. She used 2 1/5 L of orange juice, 480 ml of grapefruit juice, 800 ml of pineapple juice, and 1,200 ml of club soda. How much punch did she make?

5. Mark these numbers on the number line: **a.** 2 1/4   **b.** 1/4   **c.** 4 1/2   **d.** 5 3/4   **e.** 3 1/4
   Hint: First divide each whole-number interval into four parts (using three tick marks).

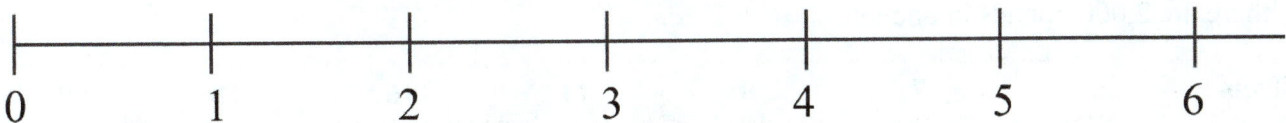

   0   1   2   3   4   5   6

6. Solve. Use *both* balances to figure out the *two* unknown shapes. Guess and check!

   44

   32

   One rectangle = _____

   One pentagon = _____

7. Draw as many different symmetry lines as you can in these shapes.

   a.

   b.

   c.   d.

85

1. A package of 12 granola bars costs $3.83.
   What will be the total cost for 48 bars?

   My estimate:

   My calculations:

2. Write an addition sentence, adding what is colored and what is not.

   a.

   b.

   c.

3. Draw an L-shape with an area of at least 24 square units.
   Write a number sentence for its area.

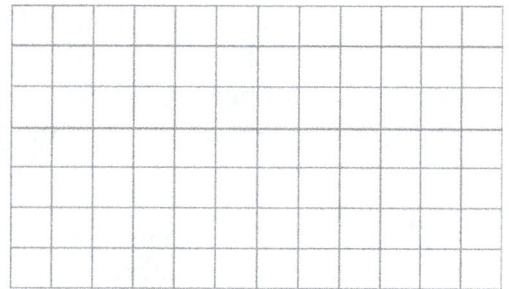

4. Fill in the table. Remember,
   there are 2,000 pounds in one ton.

Tons	5	7	9	11	16	18	20
Pounds							

5. Calculate. Give your answer using whole kilometers and meters.

   a. 9 km 400 m + 6 km 800 m

   b. 4 km 600 m + 7 1/2 km

6. Solve.

   Chris bought a refrigerator for $798 and now he has
   $1,736 left. How much money did he have originally?

   _____ + _____ = _____

   ←———— total ————→

Skills Review 81

1. Ask at least 12 people (more would be better) about their favorite day of the week.

   a. Make a frequency table and a bar graph of their answers.

Day	Frequency
Sunday	
Monday	
Tuesday	
Wednesday	
Thursday	
Friday	
Saturday	

   b. What was the most popular day of the week?

   c. Which days if any, were equally popular?

2. Three bottles of shampoo cost $4.74. Fill in the table.

Dollars			$4.74							
Bottles	1	2	3	4	5	6	7	8	9	10

3. Write these fractions as mixed numbers. Color parts to help.

   a. $\dfrac{32}{12} =$

   b. $\dfrac{18}{5} =$

   c. $\dfrac{23}{8} =$

4. Solve. Write a number sentence with an unknown.

   The perimeter of this rectangle is 42 cm. One side is 7 cm. How long is the other side?

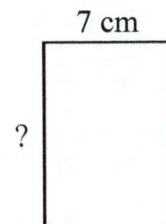

   7 cm

   ?

   _____

   Solution: _?_ = _____

Skills Review 82

1. How much is the discount or the original price?

a.

Old price $192.99

New price $175.38

Discount $_____

b.

Before $_____

Now $17.88

Discount $4.25

2. Write if each figure is a line, ray, line segment, or an angle, and name it.

a. _____

b. _____

3. Use rulers or measuring tapes to find out which is a longer distance.

   a. 5 cm or 2 inches   **b.** 7 inches or 16 cm   **c.** 12 cm or 4 inches

4. Add the fractions. Give your final answer as a whole number or mixed number if possible.

a. $\dfrac{7}{9} + \dfrac{6}{9} + \dfrac{5}{9} =$	**b.** $\dfrac{2}{5} + \dfrac{8}{5} + \dfrac{6}{5} =$
c. $\dfrac{13}{11} + \dfrac{4}{11} =$	**d.** $\dfrac{4}{6} + \dfrac{12}{6} + \dfrac{7}{6} =$

5. Divide.

a. $480 \div 6 =$ _____	**b.** $480 \div 60 =$ _____	**c.** $4,800 \div 6 =$ _____

6. Solve.

a. $960 \div 8 - 40 \times 3$
b. $60 \times 60 + 3,200 \div 4$

7. Round the number…

number	276,302
…to the nearest 1,000	
…to the nearest 10,000	
…to the nearest 100,000	

Skills Review 83

1. Subtract mentally. Compare the problems.

a. $700 - 4 =$ _____

$700 - 40 =$ _____

$700 - 400 =$ _____

$700 - 44 =$ _____

2. Divide. Mark off the problem if it is impossible to do.

a. $53 \div 0 =$ _____	**d.** $1 \div 1 =$ _____
b. $64 \div 8 =$ _____	**e.** $0 \div 1 =$ _____
c. $0 \div 18 =$ _____	**f.** $0 \div 0 =$ _____

3. Multiply.

a.
$$\begin{array}{r} 47 \\ \times\ 50 \\ \hline \end{array}$$

b.
$$\begin{array}{r} 32 \\ \times\ 76 \\ \hline \end{array}$$

c.
$$\begin{array}{r} 608 \\ \times\ 90 \\ \hline \end{array}$$

d.
$$\begin{array}{r} 529 \\ \times\ 40 \\ \hline \end{array}$$

e.
$$\begin{array}{r} 93 \\ \times\ 38 \\ \hline \end{array}$$

4. Convert between yards and feet.

a. 8 yd = _____ ft	**b.** 4 yd 2 ft = _____ ft	**c.** 27 ft = _____ yd

5. The table shows how many books Leanne read during the first six weeks of summer vacation. How many books did she read per week, on average?

Week 1	4
Week 2	0
Week 3	3
Week 4	5
Week 5	5
Week 6	7

6. Kayla drove 349 miles in a week. Mark drove four times as many miles as what Kayla drove. What is the total number of miles they both drove?

7. Add the mixed numbers. You can shade the parts to help.

a. $2\dfrac{5}{6} + 1\dfrac{4}{6} =$

b. $1\dfrac{3}{8} + 1\dfrac{7}{8} =$

Skills Review 84

1. The line graph gives the average minimum temperatures for each month in Detroit, Michigan.

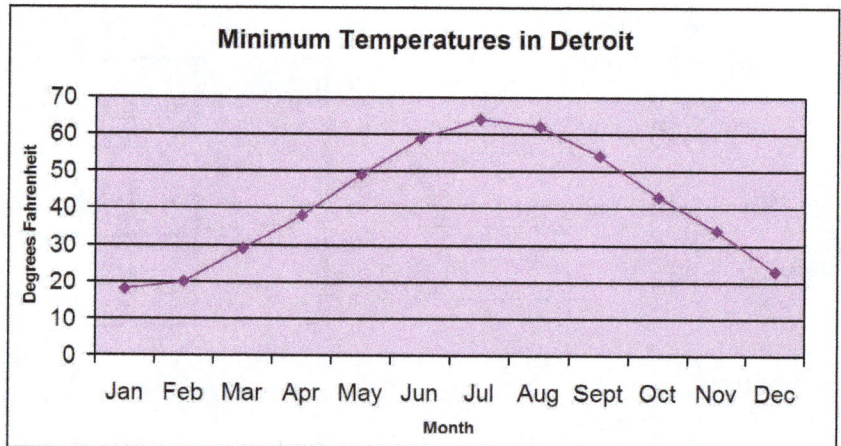

   a. Which months had an average minimum temperature higher than 50° Fahrenheit?

   b. How many degrees does the minimum temperature change from January to June?

Minimum Temperatures in Detroit

2. Write a number sentence for each word problem.

a. Allison had 75 chairs. She arranged them in rows of eight. How many full rows of chairs did she get?
b. A company hired some seven-passenger minivans to transport its 66 employees to a picnic. How many minivans did they need?

   c. In which months does the temperature typically dip below the freezing point in Detroit?

3. Compare the fractions.

   a. $\dfrac{5}{4} \square \dfrac{4}{5}$ **b.** $\dfrac{6}{8} \square \dfrac{2}{3}$ **c.** $\dfrac{10}{12} \square \dfrac{9}{6}$ **d.** $\dfrac{8}{7} \square \dfrac{5}{2}$

4. Draw four dots and connect them so that you get a quadrilateral. Measure all the angles of your quadrilateral. Then add the angle measures. Did you get 360 degrees, or close?

Skills Review 85

1. Change the times to the 24-hour clock times.

   a. 3:25 p.m. _____ : _____

   b. 9:40 p.m. _____ : _____

2. Round these numbers to the nearest ten cents and to the nearest dollar.

n	$74.29	$8.62
rounded to the nearest ten cents		
rounded to the nearest dollar		

3. Draw your own design and find its mirror image.

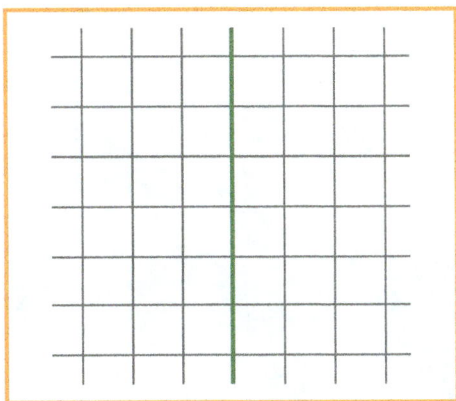

4. Draw angles for these measurements on a blank sheet of paper.

   a. 64° **b.** 49°

5. Answer the questions. You may need long division.

   a. Is 504 divisible by 7?

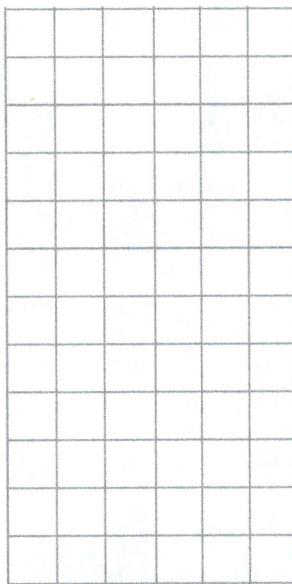

   b. Is 9 a factor of 2,748?

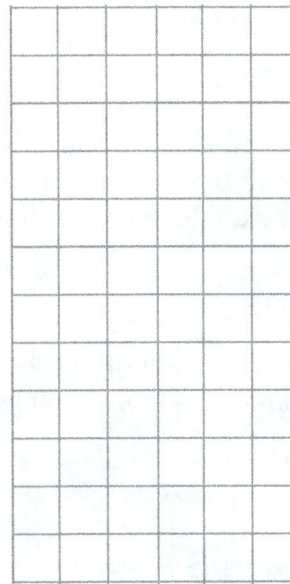

6. Maya typed a 480-word essay in 20 minutes. On average, how many words did she type per minute?

7. Write the fractions as decimals and vice versa.

a. $\frac{4}{10}$	**b.** $6\frac{7}{10}$
c. $8\frac{3}{10}$	**d.** 0.8
e. 32.6	**f.** 16.4

8. Color the first fraction. Shade the same amount of pie in the second picture. Write the second fraction.

   a. $\frac{4}{6}$ =

   b. $\frac{8}{10}$ =

Skills Review 86

1. Round the numbers to the nearest thousand, and write the **rounding error:** the difference between the number and the rounded number.

Number	Rounded number	Rounding error
12,327		
9,591		

2. Subtract. Change the mixed number into a fraction first.

a. $1\dfrac{5}{8} - \dfrac{7}{8}$ **b.** $2\dfrac{3}{12} - \dfrac{9}{12}$

3. Rachel bought 14 yards of ribbon at $3.48 a yard. *Estimate* the total cost.

4. One king cobra was 13 ft 4 in long and another one was 9 ft 8 in long. What was difference in their lengths?

5. Farmer Brown divided his 79 cows as evenly as possible between four large pastures. How many cows did he put in each pasture?

6. Write an addition for the angle measures. Use an unknown (x) for one angle measure. Then solve it.

7. Add or subtract.

a. $2.4 + 0.9 =$ _____

b. $1.5 - 0.8 =$ _____

c. $7.6 + 0.7 =$ _____

d. $3.2 - 0.4 =$ _____

8. Find the fractions that are missing from the additions.

a. $\dfrac{3}{12} +$ _____ $= 2$

b. $1\dfrac{5}{9} +$ _____ $= 3\dfrac{2}{9}$

c. $5\dfrac{4}{7} +$ _____ $= 8\dfrac{6}{7}$

Skills Review 87

1. Estimate the measures of all the angles in the triangles. Then measure with a protractor to check.

a.

_____°,

_____°, and

_____°.

b.

_____°,

_____°, and

_____°.

2. Eight pillows cost $72. How much would seven pillows cost?

3. Compare. Write $<$, $>$, or $=$ between the numbers.

   a. $0.50 \boxed{} \dfrac{1}{2}$
   b. $3.29 \boxed{} 3.92$
   c. $2.03 \boxed{} 2.3$
   d. $0.6 \boxed{} 0.06$

4. Starting at the top, find a way through the maze by coloring the equivalent fractions.

$\dfrac{2}{3}$	$\dfrac{6}{12}$	$\dfrac{1}{4}$	$\dfrac{4}{10}$
$\dfrac{3}{8}$	$\dfrac{1}{3}$	$\dfrac{1}{2}$	$\dfrac{8}{12}$
$\dfrac{2}{7}$	$\dfrac{4}{8}$	$\dfrac{7}{9}$	$\dfrac{3}{5}$
$\dfrac{2}{4}$	$\dfrac{6}{10}$	$\dfrac{4}{12}$	$\dfrac{1}{3}$
$\dfrac{3}{7}$	$\dfrac{3}{6}$	$\dfrac{5}{8}$	$\dfrac{7}{10}$
$\dfrac{4}{10}$	$\dfrac{2}{3}$	$\dfrac{5}{10}$	$\dfrac{6}{11}$

5. The temperature rises or falls. Write the new temperature.

Now	temperature rises 5°C	After
a. -8°C		_____

Now	temperature falls 8°C	After
b. 2°C		_____

Now	temperature rises 6°C	After
c. -7°C		_____

6. Solve. Give your answer as a mixed number if possible.

a. $7 \times \dfrac{2}{5} =$	**b.** $4 \times \dfrac{5}{8} =$	**c.** $6 \times \dfrac{3}{4} =$	**d.** $9 \times \dfrac{4}{7} =$

Skills Review 88

1. Add or subtract, thinking in whole thousands.

    a. $32,000 - 6,000 =$ **b.** $290,000 + 50,000 =$

    c. $780,000 + 40,000 =$ **d.** $800,000 - 80,000 =$

2. Divide. Check each result in the empty space by multiplication and addition.

a. $736 \div 9$ Check: 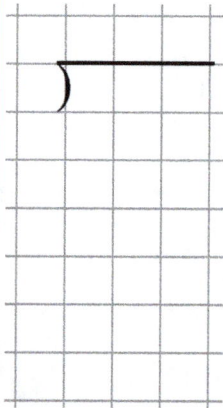	**b.** $461 \div 6$ Check: 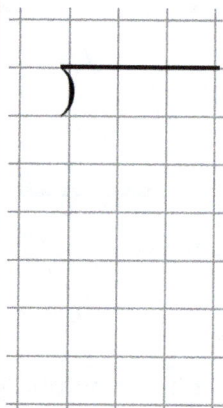

3. Which line segments in this figure are parallel? Which are perpendicular?

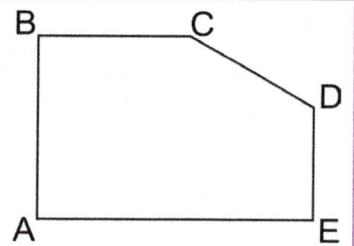

4. Subtract.

a. $5\frac{3}{8} - \frac{7}{8} =$	b. $3\frac{2}{5} - \frac{3}{5} =$	c. $7\frac{1}{10} - 4\frac{6}{10} =$

5. It took Lillian two hours and 25 minutes
to do her homework, and she finished at 5:15 p.m.
What time did she start?

6. Put operation symbols + or − into the number sentences so that they become true.

a.	b.	c.
$1.3\ \square\ 0.6\ \square\ 2.7 = 3.4$	$8.2\ \square\ 1.5\ \square\ 7.9\ \square\ 2.4 = 4.2$	$6.4\ \square\ 3.9\ \square\ 5.4 = 4.9$

Skills Review 89

1. Solve the problem.

> Sarah bought four blankets that cost $78.60 in total. Then Dana bought three of the same kind of blankets.
>
> **a.** Estimate how much Dana paid.
>
>
>
> **b.** Now calculate exactly how much Dana paid.

2. Compare.

a.	$\dfrac{3}{4}$ ☐	$\dfrac{2}{5}$
b.	$\dfrac{2}{3}$ ☐	$\dfrac{6}{6}$
c.	$\dfrac{5}{8}$ ☐	$\dfrac{3}{8}$
d.	$4\dfrac{6}{10}$ ☐	$4\dfrac{5}{12}$
e.	$\dfrac{4}{6}$ ☐	$\dfrac{1}{2}$

3. These students practiced finding factors. Be a teacher detective, and check and correct their work.

a. Alex found all the factors of 56:	**b.** Brenna found all the factors of 42:
$56 = 1 \times 56$ $56 = 2 \times 28$ $56 = 4 \times 15$	$42 = 1 \times 42$ $42 = 2 \times 21$ $42 = 4 \times 11$
$56 = 8 \times 7$ $56 = 7 \times 8$	$42 = 6 \times 8$ $42 = 7 \times 6$
The factors are 1, 2, 4, 7, 8, 15, 28, 56	The factors are 1, 2, 4, 6, 7, 8, 11, and 21.

4. There are six people in Rachel's family, and they each had 3/4 cup of cereal for breakfast. How much cereal did they eat in total?

5. Add or subtract in columns. Line up the decimal points.

a. $64.73 + 29.5$	**b.** $35.94 - 16.28$	**c.** $29 + 8.63 + 43.7$

Skills Review 90

1. Convert between kilograms and grams. Use decimals.

a. 800 g = _____ kg

b. 0.9 kg = _____ g

c. 40,700 g = _____ kg

d. 9,200 g = _____ kg

e. 32.6 kg = _____ g

2. Multiply. Estimate the answer on the line.

a. 78×45	**b.** 326×9
Estimate: _____ × _____	Estimate: _____ × _____
= _____	= _____

3. Draw any parallelogram.

4. Label the triangles in the pictures as right, acute, or obtuse.

a.   **b.**   **c.**   **d.**

5. Write the temperature shown by the thermometer. Then describe a situation to fit that temperature.

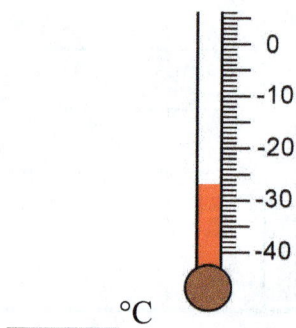

0
-10
-20
-30
-40

_____ °C

6. Starting at the top, find your way through the maze by coloring the numbers that are primes.

12	8	21	29	33
16	13	41	27	81
45	19	18	146	94
132	73	107	181	215
64	75	121	163	112

www.ingramcontent.com/pod-product-compliance
Lightning Source LLC
Chambersburg PA
CBHW051228200326
41519CB00025B/7290